AF413660

# Renewables or Nuclear

E. E. Lewis

# Renewables or Nuclear

## Which Should Lead in Curbing Climate Change?

 Springer

E. E. Lewis
Department of Mechanical Engineering
Northwestern University
Evanston, IL, USA

ISBN 978-3-032-08073-8        ISBN 978-3-032-08074-5    (eBook)
https://doi.org/10.1007/978-3-032-08074-5



This Springer imprint is published by the registered company Springer Nature Switzerland AG
The registered company address is: Gewerbestrasse 11, 6330 Cham, Switzerland

If disposing of this product, please recycle the paper.

*In Memory of*
*Ann Flinspach Lewis*

# Preface

Containing climate change is on the forefront of challenges we face. Meeting it will require difficult decisions—and the most consequential may well be how we choose to generate electricity in a decarbonized world. Arguably, the most continuous issue faced is whether renewable or nuclear energy should play the larger role in reducing greenhouse gas emissions. The public is polarized, as indicated by nearly 40,000 members participating in the Facebook group "Renewables vs. Nuclear: DEBATE". Of books intended for the lay reader, several include "renewable" or "nuclear"—but not both—in their titles and deal with only one in depth and either ignore or disparage the other. In this volume, I strive to treat the facts—whether they favor or disfavor one or the other—in an even-handed manner and to draw conclusions from where they lead.

I've come to writing *Renewables or Nuclear* from a long career of research and teaching at Northwestern University. For the most part, my work has centered about the physics and safety of nuclear systems. My fascination with nuclear energy's possibilities stemmed from long before the term renewable energy came into use, or the threat of climate change was understood outside a small cadre of climate scientists. My interest was galvanized by the famed "Atoms for Peace" speech that President Eisenhower delivered before the United Nations in 1957. His presentation promoted the use of nuclear energy for electricity generation, agriculture, and medicine.

This book's genesis is a fireside talk I gave at Northwestern's Shepard Residential College. Preparing for it focused my attention on renewables, how they worked, their strengths and limitations, and on the developing competition between renewable and nuclear energy. A unifying theme became how wind and solar generation would interact with nuclear plants on electric power grids, and on how our economy could best be decarbonized. My study continued, visiting wind and solar farms, and intensified during pandemic-induced isolation. In August of 2020, I gave a zoom presentation to the Northwestern Emeritus Organization, which was well received by attendees from diverse fields: economics, law, medicine, journalism, of course engineering, and others. Encouragement following further talks led to my decision to write a book about the issues relating renewable and nuclear energy that would be easily understood by the intelligent layperson.

Northwestern colleagues and students, family, friends, and my wife have offered encouragement and much more. They have edited, proofread, fact checked, jogged my memory of long-ago experiences, argued opposing views, and offered suggestions to improve the narrative and make it more accessible to the lay reader. They included David Abramson, Bruce Bingman, Roger Blomquist, Roger Boye, Nickolas and Marcy Collins, James Edwards, Lloyd and Mary Evans, Jeffrey Garrett, Ann Lewis, Paul and Gloria Lewis, Sohrab Mobarhan, Firouz Niazi, two generations of the Pardoe family (Elizabeth Jeremy, Frederick, and Robert), Mark Seniw, Daniel Taber and Tom Tamm. I greatly appreciate the efforts of my editors Michael McCabe, Brian Halm, Meenah Kumary, and Avanthika Babu at Springer Nature, for they have made the book much better than it otherwise would have been. The book's dedication is to the memory of the love of my life, my wife for 60 years, Ann Flinspach Lewis.

Evanston, IL, USA                                                        Elmer E. Lewis

**Competing Interests** The author has no competing interests to declare that are relevant to the content of this manuscript.

# Contents

# Chapter 1
# Our Fossil Fueled Legacy

**Keywords** Fossil fuels · Industrial revolution · Greenhouse gases · Pollution · United Nations · Paris Accords · Electricity · Decarbonization · Carbon dioxide · Economic sectors

The first age of renewable energy was coming to an end in seventeenth-century England.[1] The population was growing, and nascent industries were expanding. Wood was essential for space heating and to fire lime kilns and iron forges. The demand for its bioenergy to produce charcoal for the smelting of iron was insatiable. Trees suitable for cropping could not be regrown fast enough to meet demand. Forests were denuded, and firewood was imported from Scandinavia as the energy crisis grew. Rapidly expanding the small-scale mining of coal became the solution. Coal was far cheaper than wood, and its economy brought expansion of existing industries and the creation of new ones. Thus humankind began digging into its inheritance of nonrenewable fossil fuels.

The mid-eighteenth century brought the invention of the steam engine. Powered by coal, it converted heat to mechanical energy. First used for pumping water out of coal mines, the engines soon superseded the use of waterwheels for power production. Unlike waterwheels, steam engines were neither geographically restricted, nor robbed of their power by seasonal variations in the flows of rivers and streams. These coal-fired engines facilitated the mechanization of the textile and other handicraft industries. Subsequently, manufacture of lighter engines allowed them to be put on a rail car—followed by a coal car—or on a ship. By the end of the eighteenth century, railroads were expanding, followed by steamships a few decades later.

Powered by coal, the industrial revolution spread outward from Britain, eventually lifting living standards around the world. Industrialization spread abroad slowly where lower population densities and more extensive forests allowed biofuels to remain the primary source of heat for longer. In the nineteenth century, Germany

---

[1] https://www.economist.com/2024/12/19/how-premodern-energy-shaped-britain

© The Author(s), under exclusive license to Springer Nature Switzerland AG 2026
E. E. Lewis, *Renewables or Nuclear*,
https://doi.org/10.1007/978-3-032-08074-5_1

then other European counties converted to coal. The US conversion was slower, where in 1850, more heat was still being produced from firewood.

Coal production continued to grow worldwide. As shown in Fig. 1.1, it remained the dominant energy source until well after the 1859 discovery of oil in Pennsylvania. During the last half of the twentieth century, however, the rapid expansion of automotive travel, demand for cleaner home heating, railroad conversion from coal to oil, plastics manufacture, and a host of other applications caused oil and gas to account for a rapidly increasing fraction of fossil fuel production. As Fig. 1.1 indicates, fossil fuel use continues to skyrocket.

Long before fossil fuel combustion had risen to levels causing atmospheric greenhouse gases to accumulate to dangerous levels, air and water pollution from burning coal presented a challenge. It dates back to the seventeenth century, and possibly earlier. In Britain, there were ordinances to restrict the burning of coal as the air over London became increasing smoggy. Smog was the concoction created when coal smoke was combined with moisture in the air. And the English climate was particularly prone to its formation.

In the winter of 1952–1953, coal smoke combined with stagnant moist air over London to produce a killer smog that lasted several weeks. The number of deaths from respiratory problems was horrendous. The immediate death toll was 4000, but many more died in the months that followed, leading to an estimated 12,000 deaths.[2] The twentieth-century ascendance of the automotive travel across the world caused

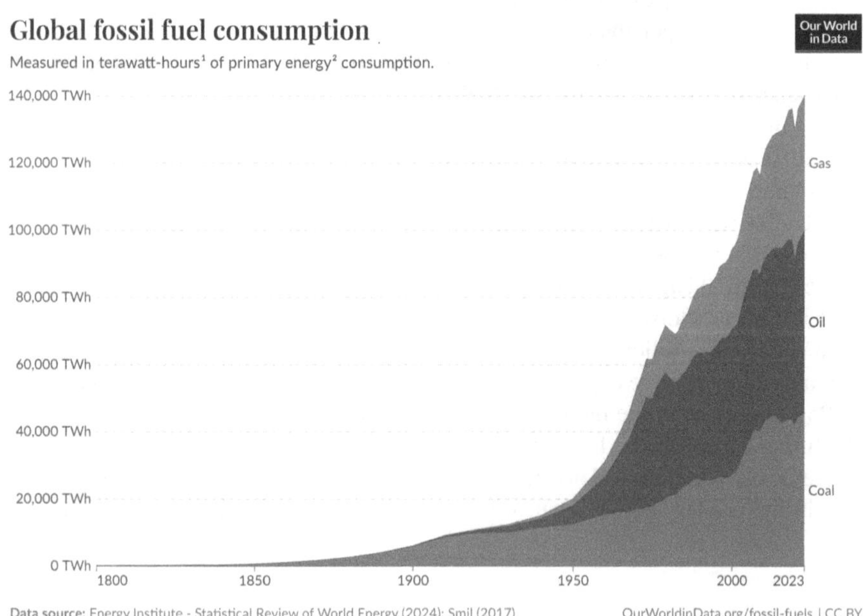


**Fig. 1.1** Global fossil fuel consumption. (Source: OurWorldinData)

---

[2] https://education.nationalgeographic.org/resource/great-smog-1952/

the combustion of gasoline, diesel fuel, and other petroleum products to add to coal smoke in blanketing major metropolitan areas, such as Los Angeles and New York in the USA, with dangerous levels of air pollution.

Pollution comes from coal's many impurities—sulfur, mercury, arsenic, and more—and from the combustion processes in all fossil fuels. Incomplete combustion produces carbon monoxide, and high-temperature combustion yields nitrous oxides. With either coal or oil, the deadliest pollutant is tiny particulate, which enters the lungs, causing cardiorespiratory disease. Beginning in the 1970s, scrubbers were required in coal plant smokestacks, and catalytic converters were mandated on motor vehicle exhaust pipes. These technologies can capture well over 90% of pollutants and have come a long way in improving the air quality in major cities of the industrialized world. In developing countries, however, air pollution is an increasing problem.

***

Glaciers melting, sea level rising, disastrous wildfires, record-breaking flooding, and more now force the public's attention to focus on another legacy of fossil fuels: climate change. While opinions on every aspect of climate change still vary widely, consensus has been reached that its cause is the atmospheric buildup of carbon dioxide and other gases produced from the combustion of fossil fuels. That buildup entraps heat in the atmosphere, creating a greenhouse effect that warms the earth's surface. For too long, the public remained ignorant of the gravity of the oncoming climate crisis. In the 1990s, however, that began to change under the auspices of the United Nations. The first assessment by the Intergovernmental Panel on Climate Change—or IPCC—was published in 1990. The first of a numbered series of the annual Conference of Parties—or COP1—meet in 1995, and COP30 recently met.

A pivotal year was 2015, when at a Paris conference, representatives from across the globe came together in recognition of climate change's dangers. The result was the Paris Climate Accords of 2015, signed by nearly 200 nations. The Accords set an ambitious goal of limiting global warming to well below 2.0 °C (3.6 °F) of preindustrial levels and preferably limit the increase to 1.5 °C (2.7 °F). They specify that emissions should be reduced as rapidly as possible to reach net-zero greenhouse gas emissions by 2050. Meeting that goal will be a Herculean task, for fossil fuels still supply more than 80% of the world's energy.[3] Of all nations, China emits by far the most greenhouse gases, but on a per capita basis, the USA emits the most.[4] In the USA, 14 tons of carbon dioxide ($CO_2$) per person are emitted annually, and driving a midsized car from coast to coast would emit somewhat more than a ton of $CO_2$.[5]

In the USA, the Biden administration set fourth an aggressive policy aimed at reducing greenhouse gas emissions. Executive Order 14057, *The Federal Sustainability Plan*, stipulates an ambitious path to achieve net-zero emissions across Federal operations by 2050 and mandates the use of only carbon-free

---

[3] https://www.eesi.org/topics/fossil-fuels/description

[4] https://www.wri.org/insights/interactive-chart-shows-changes-worlds-top-10-emitters

[5] https://www.statista.com/statistics/1049662/fossil-us-carbon-dioxide-emissions-per-person/

electricity by 2035. Large sums of money have been allocated toward achieving those goals in legislation known as the Inflation Reduction Act. The task is daunting. It includes making policy decisions on what methods for fighting climate change should be accentuated and which should receive less emphasis. Should renewable or nuclear energy play the larger role in eliminating greenhouse gas emissions from fossil fuel combustion?

***

Carbon dioxide is the dominant greenhouse gas, and it remains in the atmosphere for centuries. Methane, along with some other gases, contributes smaller amounts. Fortunately, methane remains in the atmosphere for only a few decades. Figure 1.2 shows the growth of carbon dioxide emission and from where around the globe it enters the atmosphere. Which fossil fuel dominates a region has a strong effect how much $CO_2$ is produced. Coal emits the most $CO_2$ per unit of energy produced, while the variety of oil-based fuels produces only 70–80% of what coal does. These range from jet fuel to container ship bunker oil, with gasoline the most widely used. Natural gas, that is methane, emits only roughly half as much $CO_2$ as coal for the same energy produced.

Carbon emissions of the industrially developed countries, most notably in Europe and the USA, have leveled off, and in some cases declined slightly. Two factors come into play. First, unlike China, India, and other developing counties, in developed countries, physical infrastructure is already in place, and their populations are growing slowly if at all, and in some nations declining. Consequently, less of their

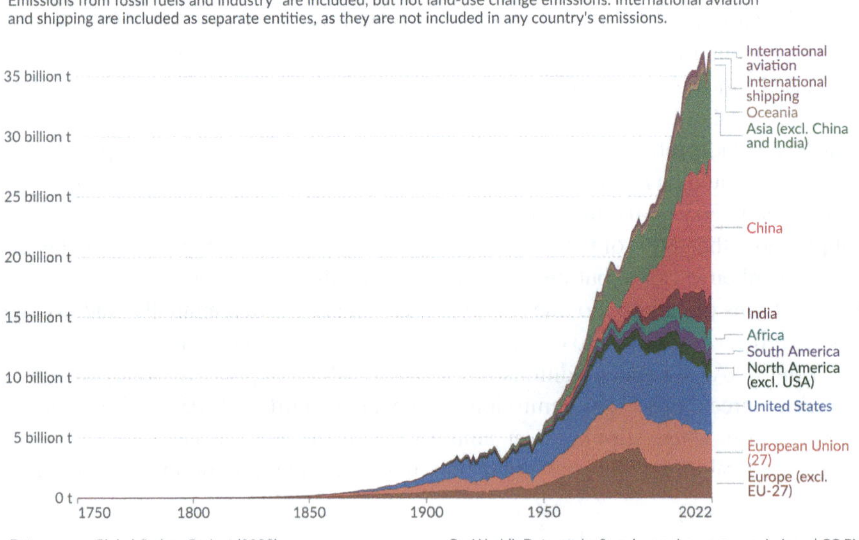


**Fig. 1.2**  Annual $CO_2$ emissions by region. (Source: OurWorldinData)

economies are devoted to heavy construction of highways, bridges, factories, power, water purification, and sewage treatment plants. Construction of these and other facilities is heavily reliant on the production of concrete and steel, two commodities whose manufacture results in emission of large quantities of $CO_2$. Second, in the USA and elsewhere, fracking and horizontal drilling technologies have brought down the cost of natural gas to where much of power production and space heating has been switched from coal or oil to gas. Since methane combustion emits substantially less $CO_2$ than coal or oil, the buildup of greenhouse gas has slowed.

Understanding how greenhouse gas emissions may be curbed requires first examining the economic sectors from which they come. These are shown in Fig. 1.3. The pie chart on the left is for worldwide emissions, and the one on the right is for the USA, used as an example of a developed country. Note the similarities and difference in sector sizes between the two charts. Electricity occupies about a quarter of each of the charts. Heat is included as a separate category on the global chart because in some counties the heat from electric power plants is used for space heating of building complexes, desalination, or for other purpose. The fractions of emissions due to industry are similar in both charts. Globally, the transportation sector is only half that in the USA. This is a consequence in part from the US dependence on personal automobiles that exceeds most other highly developed nations. Global agriculture and forestry worldwide are more than double that of agriculture in the USA. The difference may be due to the rapid cutting of the tropical forests in Brazil, Indonesia, and elsewhere to gain additional crop land. To make matters worse, burning is often used to dispose of the cut timber. The buildings and other energy categories on the Global Chart are somewhat comparable to commercial and residential categories on the US chart, with some of the differences being due, no doubt, to the differences in record keeping between the many nations included in the global results.

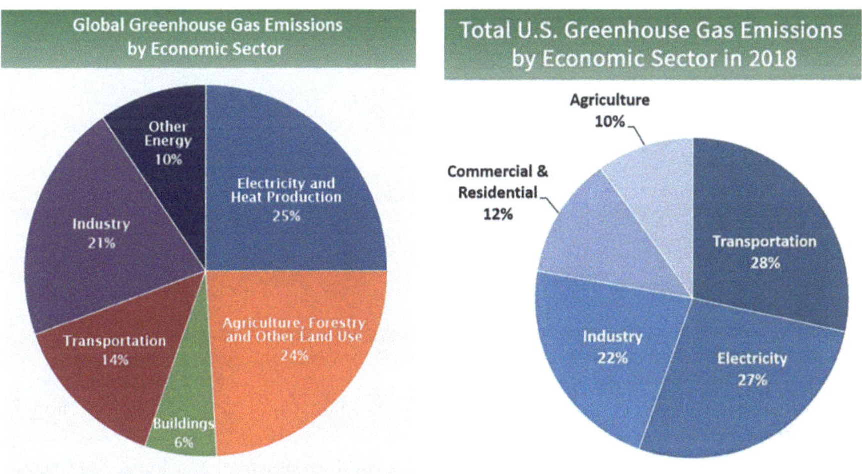


**Fig. 1.3** Annual greenhouse gas emissions by economic sector. (Source: EPA)

How can the emissions of greenhouse gases be greatly reduced or totally eliminated to meet the goal of carbon neutrality by 2050? For two reasons, consensus is that priority must be the decarbonization of electricity generation. First, we know how this can be done: renewable and nuclear energy provide means for accomplishing the task. Second, much of the carbon emissions from other economic sectors can then be eliminated by converting their power sources to decarbonized electricity. But the demand for electricity will be huge, accelerated by electricity consumption for data centers, electric automobiles, and other advancing technologies.[6] Electricity demand is expected to nearly quadruple by the time the goal of carbon neutrality is reached.[7]

Most of transportation presently relies on petroleum-powered vehicles. Much progress has already been made in producing cars, trucks, and buses powered instead by electric batteries. Hydrogen produced from decarbonized—often called "green"—electric generators can power fuel cells. These offer attractive prospects for decarbonizing shipment by large trucks, railroads, and ships. In some countries, railroads have been powered directly by electricity for many decades.

Most of the greenhouse gas emissions from commercial and residential buildings result from space heating, which can be electrified, most efficiently by the use of heat pumps. Many of the processes that constitute the industry sector require electricity as well as high-temperature heat. In some cases, that heat may be provided from the waste heat generated by nuclear power or biomass production. Though less efficient when the needed temperatures are too high to use waste heat, electricity can be converted to higher-temperature heat. Agriculture is perhaps the most difficult sector to decarbonize completely, although electrifying tractors and other farm equipment as well as the manufacture of fertilizer would result in a sizable reduction in emissions.

---

[6] https://www.wsj.com/business/energy-oil/big-techs-latest-obsession-is-finding-enough-energy-f00055b2

[7] E. Larson et al, Net-Zero America: Potential Pathways, Infrastructure, and Impacts, Princeton U. (2021).

# Chapter 2
# Grid Structure, Stability, and Balance

**Keywords** Electric grid · Blackout · Grid stability · Turbine-generator ·
Synchronous inertia supply and demand · Reliability · Baseload power · Load
following · Thermal shock · Long term planning

On the afternoon of June 27, 1971, the temperature in Chicago reached a record
breaking 101 °F. Accelerated air conditioning use caused the demand for electricity
to reach a record level. At Commonwealth Edison's dispatcher's complex in the
city's Loop, the tension was palpable, felt by everyone from senior management to
junior engineers. Electricity demand must be supplied exactly on a second-by-
second basis, for it could not be stored in significant amount. If supply were to fall
short of demand—also called the load[1]—even for only several seconds, generators
would shut down automatically to prevent damage. In professional parlance, the
load would be lost, and a cascading blackout would spread with lightning speed,
enveloping the Chicago metropolitan area and a large swath of Northern Illinois.
Hours if not days would be required to restore power.

On one wall, charts recorded the electrical output of each of Edison's power
plants: coal, oil, gas, and nuclear. On another, a large map showed Edison's grid as
a network of lighted points, indicating power plants and substations, with lines
between them delineating high-voltage transmission lines. Colored coded lights
indicated which lines were in service. In front of the dispatchers, the Economic
Generation Control System automatically determined which generators should be
running, choosing to add the most economical available at any given time. At that
time, every generator was running full bore.

While data was displayed on dials, charts, and the map, frequent phone conversa-
tions between the dispatcher's complex and personnel at the power plants kept close
tabs on idiosyncratic behaviors—problems that might be developing and more.
Edison couldn't afford for anything to go wrong at that time of afternoon, for the
demand from air conditioner use would suddenly spike. Many new air conditioner

---

[1] Load and demand are used interchangeably in the chapters that follow.


E. E. Lewis, *Renewables or Nuclear*,
https://doi.org/10.1007/978-3-032-08074-5_2

owners would wait till the room felt hot before turning on their recently purchased devices. The net result would be a sudden increase in demand seen in the control room.

Attention focused on Dresden 2, a nuclear reactor with boiling water coolant, because it had only recently come online. My concern was with the insertion lengths of the neutron absorbing control rods, and the speeds of the pumps circulating coolant through the reactor's core. I had good reason, for my consulting had involved me in determining a safety factor required by the Nuclear Regulatory Commission: the cold shutdown margin. It was the margin by which the neutron chain reaction would be shut down—taken subcritical in nuclear jargon. That margin had to be maintained even after the reactor's power was reduced to zero and its core cooled to room temperature, that is to "cold." But the margin could not be measured directly while the reactor was operating. Instead, it had to be calculated in terms of two variables I had been watching. Both changed only slowly over hours, days, and weeks as the uranium fuel was depleted. Though slow, those changes added to the complexity of the calculations.

The calculations were imprecise, performed on late 1960s main frame computers, which occupied large rooms but were less powerful than today's laptops. I dreaded the possibility that a combination of control rod insertions and pump speeds would arise indicating that the safety factor had been violated. Based on calculations of questionable precision, shutdown could be forced on a reactor that seemed to be operating satisfactorily. Thankfully, the control rod insertions and the circulation pump's speeds did not reach a combination indicating the safety criterion had been breached.

A problem causing the 894 Megawatt (MW) Dresden 2 reactor or any one of the utility's power plants to shutdown would have caused supply to fall short of the record high load, making it unlikely that the system could recover sufficiently to avoid a blackout. Even with all the power plants running flawlessly, the threat of a blackout couldn't be ignored. A further increase in demand could surpass the utility's capability of meeting it. Options were limited. Chicago Steel and Wire already had shut down in accordance with an agreement with Commonwealth Edison to do so during hours of high demand in exchange for lower electric rates. On that day, no other companies had such agreements. To make matters worse, the heat wave was widespread across the middle west. Thus, neighboring utilities had no spare power to send through transmission line interconnects.

Grabbing everyone's attention was the display panel indicating the frequency of alternating current circulating through the utility's system of power plants, high-voltage transmission lines, substations, and distribution networks. The display needed to read exactly 60.0 cycles per second (i.e., 60 Hertz or Hz). If the power supplied fell short of demand, the frequency would droop, and if it fell much below 59.7 Hz, after a few seconds generators would start shedding load to prevent them from being damaged. A cascading blackout would bring down the entire Commonwealth Edison system, leaving millions of people in darkness.

A long blackout would do a great deal of harm beyond the lights going out. Elevators would stop; traffic signals go dark, vital machinery would stall or shut down, and over time, emergency generators would run low on fuel. The heat would become unbearable, particularly in new buildings designed with the assumption that air conditioning would always be available. The 1965 blackout that spread over the northeastern states and left 30 million people in darkness, many of them for hours, remained fresh in the engineers' minds.

Fortunately, the rotational inertia of the massive spinning turbine-generators at Edison's several power plants would slow the frequency decline, giving the dispatchers precious time to take critical actions. The first would be to make preplanned reductions in voltage, thereby reducing electricity usage. Lights would dim slightly, but the reductions would be small enough that motors driving air conditioners, refrigerators, elevators and other equipment would not stall. If the voltage reductions were not sufficient to stabilize the frequency, then rolling blackouts would be initiated, successively depriving the residents of various geographical areas of power for a few hours at a time.

The frequency edged below 60.0 Hz, as the record-breaking electricity use peaked, but then slowly returned to 60.0 Hz as demand subsided. Tension eased in the control room as the outside temperature slowly decreased and the power demand followed suit. Thankfully the Dresden 2 reactor and the other power plants withstood the stress test, and power supply was able to match demand second-by-second that hot afternoon. The load was not lost!

***

Much is changing in the transition between electricity supplied predominately by fossil fuel combustion and a future based on low-carbon electricity from renewable and nuclear energy. Grids, however, will endure, along with the fear of losing the load, for they are unique in accommodating the fundamental properties of electricity. Electricity transmits at nearly the speed of light between distant points of generation and consumption, and presently it cannot be stored in adequate amounts or for long enough times to substantially impact grid operations. Grids allow electricity to be generated, transmitted, and distributed instantaneously and efficiently.

Figure 2.1 shows a greatly simplified diagram of such a grid. Electricity is generated on the left as indicted by icons representing nuclear, solar, fossil fueled, and wind power plants. The electricity is increased in voltage at substations along the vertical line between generation and transmission. A network of high-voltage transmission lines, depicted simply by four transmission towers, delivers the electricity to a second set of substations along the vertical line between transmission and distribution. Those substations reduce its voltage, for a network of lower voltage lines to deliver power to the residential, commercial and industrial consumers shown as icons on the right.

The icons on the left of Fig. 2.1 portray a combination of nuclear, fossil-fueled and renewable power plants. Before the introduction of renewable and nuclear energy, all the power plants would have been fossil fueled, except where hydroelectric or geothermal generation could contribute significantly. At present, the mix of power sources is evolving rapidly, led by the increase in wind and solar generation. If decarbonization goals are to be met, grids will be dominated by renewable and nuclear energy. In reality, electric grids contain many substations, powered by

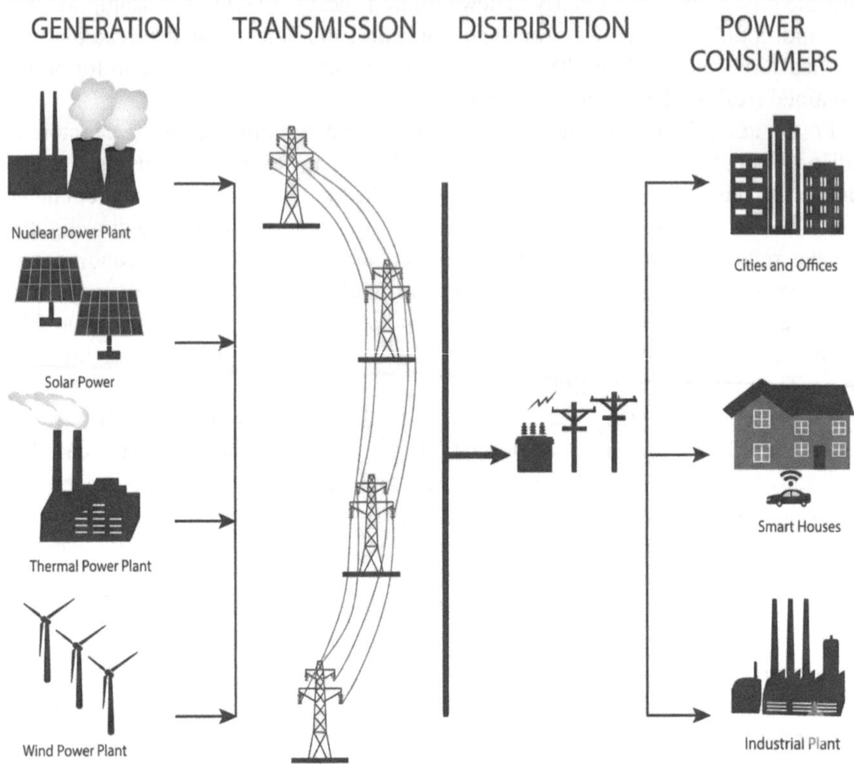


**Fig. 2.1** Electric grid diagram. (Source: Shutterstock)

multiple power plants, often numbered in the dozens. They contain complex networks of high-voltage transmission lines. All grid components are synchronized to a single frequency, 60 Hz in the USA but 50 Hz in many other nations.

Ten grids span the USA, each powering a geographic area indicated by color in Fig. 2.2. Each grid's transmission lines are connected to and synchronized with its geographical neighbors, allowing power to be shared. Such sharing becomes indispensable when widespread adverse weather, equipment failures, or other difficulties cause power on a grid to be in short supply. All the grids interconnected through such arrangements must be synchronized with each other to 60 Hz. Taken together, they are termed synchronous networks or interconnections.

The 3000 power plants of the continental USA provide a million megawatts (MW) of generating capacity to a network of more than 200,000 miles of high-voltage lines and 5.5 million miles of local distribution lines.[2] The National Academy of Engineering ranks this accomplishment as the greatest engineering achievement

---

[2] EPRI, The Green Grid: Energy Savings and Carbon Emission Reductions Enabled by a Smart Grid, Technical Update, June 2008.

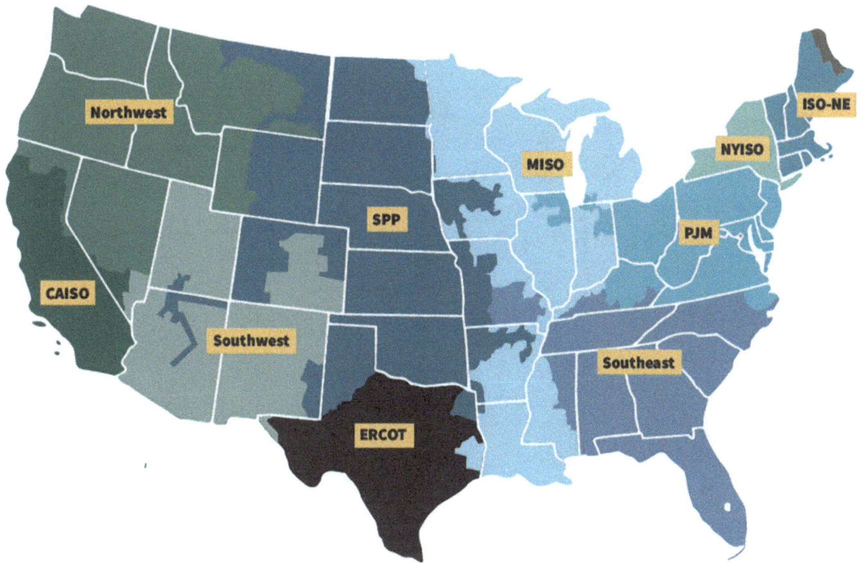



**Fig. 2.2**  Grids in the continental U.S. (Source: FERC)

of the twentieth century. The grid designated by the acronym PJM is the largest
provider of power.[3] It spans part or all of 13 states and Washington DC and spreads
from the Atlantic coast to the Mississippi river with a network consisting of
85,000 miles of high-voltage lines. Figure 2.3 shows its high-voltage transmission
lines and substations. The diagram's line colors indicate various high-voltage levels,
and the dots indicate substations.

The ten US grids are federated into three larger synchronous networks, called
interconnections. The continental USA is divided into the Eastern, the Western, and
the Texas interconnections. The boundary between Eastern and Western
Interconnections is the Rocky Mountains. Alternating current cannot be shared
between interconnections, for although they are all synchronized to 60 Hz, the three
are not synchronized with each other. At several locations along the border between
the Eastern and Western Interconnections, alternating current can be rectified to
direct current, passed through the interface, and converted back to alternating current. The amount of power that can be transferred, however, is quite limited. The
Texas Interconnection is isolated from the other two to avoid federal regulations
related to interstate commerce.

Across the world, electricity is generated, transmitted, and distributed through
grids, often linked together as in the USA to form synchronous networks. These are
shown in Fig. 2.4. Within each synchronous network, all power plants produce

---

[3] Stands for Pennsylvania–New Jersey–Maryland, the three states within which the original 1927
power pool operated.

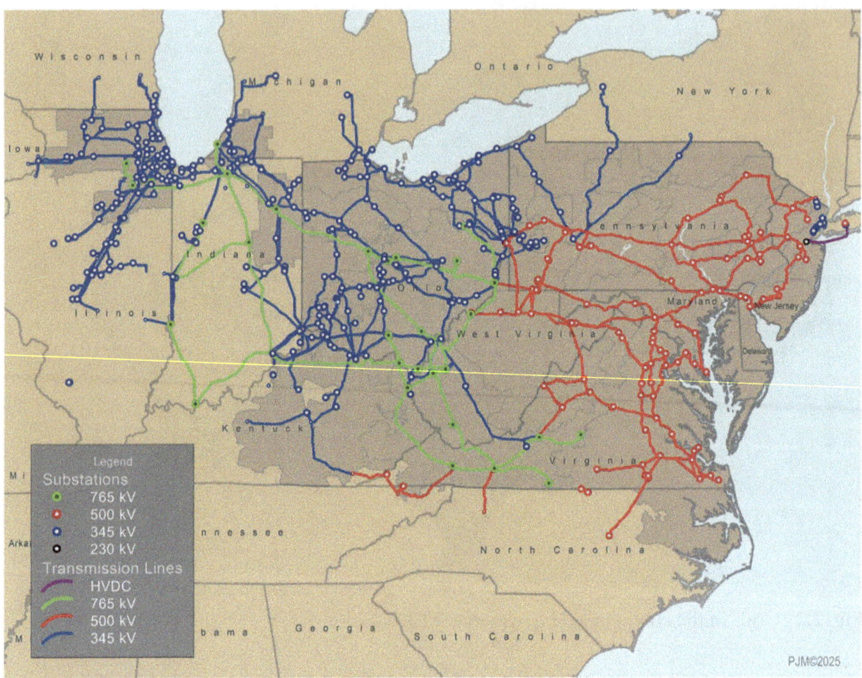



**Fig. 2.3** PJM high voltage lines. (Source: PJM)

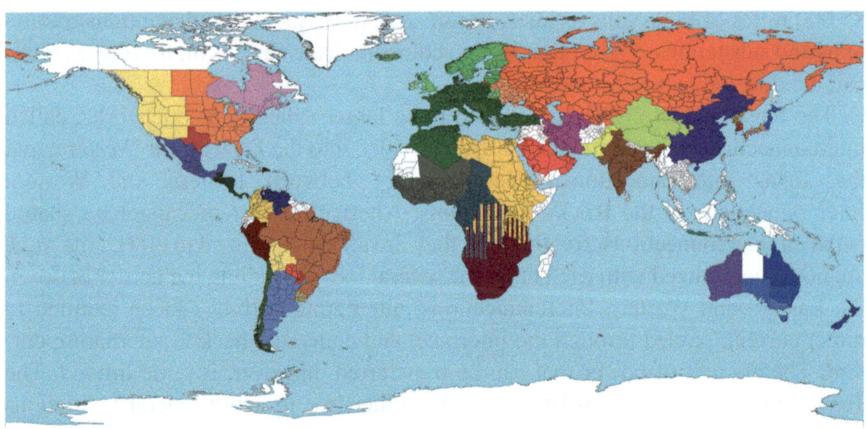

**Fig. 2.4** Global synchronous grids. (Source: Commons, Wikimedia.org)

alternating current synchronized exactly with each other to a single frequency. In North America, it is 60 Hz. In South America, there are both 50 Hz and 60 Hz networks. With rare exception, such networks in the Eastern Hemisphere are 50 Hz; Japan, however, has both 50 and 60 Hz networks. Because the colored synchronous networks of Fig. 2.4 are not synchronized with each other, to transmit electricity between them, it must first be converted from alternating to direct current, cross the border, and then be converted back to alternating current synchronized to the network it is entering.

The world's largest synchronous network spans continental Europe, serving 600 million customers, living in 24 countries. It generates nearly 100 thousand megawatts (MW) of 50 Hz power.[4] Its 50 Hz power is connected through direct current interfaces to three adjoining synchronous networks: the Nordic, the Baltic, and the British Isles. Within these networks, there are 39 transmission systems operators in charge of grids that somewhat resemble those that make up the three US synchronous networks. Together, they are titled the European Network of Transmission System Operators for Electricity or ENTO-E. In several nations, such as France, there is a single synchronous grid, while in other nations there are a variety of configurations; Germany has four synchronous networks, each consisting of several grids. India's single synchronous network consists of five grids,[5] and Japan has 4 synchronous networks, three 60 Hz and one 50 Hz, powered by 10 grids.[6]

Operating a grid, large or small, is an arduous task, since electricity cannot be stored economically in significant amounts. Thus, second by second supply must exactly equal demand in all kinds of weather, even when generators, transmission lines or other equipment fails. The supply-demand balance must be accomplished without violating changing governmental regulations, or environmental constrains. Growth of demand must be anticipated years in advance, for new power plants, transmission lines and substations must be planned, built and up and running in time to adequately supply the growing demand. And all this must be accomplished while keeping the grid reliable and the price of electricity at acceptable levels.

In a properly functioning grid, frequency across the entire system must be 60 Hz or 50 Hz. Voltage takes on different values at different grid locations, for example, at opposite ends of high-voltage transmission lines and at opposite sides of step-up and step-down substations. At any one grid location, however, voltage should not change, except under the control of the grid-balancing authority. Currents are what change in transporting energy between grid locations. For the grid to meet the foregoing conditions, it must be stable and balanced. Stability is the grid's ability to recover from system disturbances without a blackout resulting, no matter whether the disturbance arises from a failure of grid equipment or from an external cause. Balance means the supply is always equal to demand at every location on the grid and no matter what the conditions under which the grid is operating. We first examine stability and then, in the subsequent section, balancing.

***

Turbine generators are the predominant source of grid stability, acting as grid shock absorbers. They are powered by hydro, nuclear, geothermal or fossil fuel

---

[4] https://www.dgplusdesign.com/insights/inside-the-worlds-largest-interconnected-grid

[5] https://www.climatescorecard.org/2023/01/indias-national-power-grid/

[6] https://www.jonesday.com/en/insights/2015/08/japan-approves-final-set-of-power-market-reforms

energy. Among the largest are the steam turbine generators of nuclear plants, some producing substantially more than 1000 MW of electricity. Like the one shown in Fig. 2.5, they contain hundreds of tons of steel spinning always at 60 Hz, no matter whether they are operating at partial or full power. Depending on turbine type, the rate at which steam, water, or gas enters their intake determines their power level. The synchronous inertia of the massive amounts of spinning steel acts as a buffer against rapid change. This inertia stabilizes the grid against disturbances that may come from many causes.

For an analogy to a turbine, think of a bicycle. At high speeds, the mass of its spinning wheels provides enough rotational inertia to keep the bike quite stable, dampening disturbances caused by bumpy roads or gusty winds. On racing bikes with lighter wheels, there is less rotational inertia, and maintaining stability requires more effort. Likewise, at slower speeds the decreased rotational inertia will reduce stability, meaning more effort is needed to prevent tipping. At rest, of course, a bicycle has no stability. In contrast, however, the rotation speed—and thus the inertia—of a turbine generator always remains at 60 (or 50) Hz, no matter how much electricity it is producing.

Power quality is necessary for electric equipment, particularly computers and other digital devices, to function properly. Frequency regulation and voltage support are its two primary aspects. If demand exceeds supply, frequency will droop. Frequency stability is paramount in maintaining power quality, for if it deviates from 60 Hz by much more than half of a percent—more than 0.3 Hz—for even a few seconds, generators and other equipment will begin detaching from the grid to prevent damage, thus initiating a cascading blackout. Voltages are allowed somewhat more variation than frequency, typically within 5% of the target value. Larger

**Fig. 2.5**  Large turbine-generator. (Source: NRC)

deviations are allowed for very short times, for example, 10% excess voltage for less than 0.5 s or 10% low voltage for less than 10 s.

Over timespans of only seconds, demand undergoes many small random fluctuations caused by many residential, commercial and industrial customers turning on and off, lights, heat, air conditioning, appliances and a plethora of other electrical devices. The synchronous inertia of turbines' spinning masses damps such fluctuations quickly, stabilizing frequency and returning voltages to their target values.

In addition to supplying synchronous inertia for grid stabilization, turbine generators must also be tuned to provide a required amount of reactive power. Few, others than electrical engineers working in the power industry, have ever heard of it. Instead of reading the following paragraph, returning to the bicycle analogy may be helpful. Most of the rider's energy is spent peddling the bicycle forward, but a smaller amount must be devoted to keeping it from tipping—to keeping it stable. The former is analogous to real, the latter to reactive power. The reactive power doesn't propel the bike forward (i.e., it does no work) but is necessary to keep from tipping over. Too much reactive energy is wasteful of the rider's limited total energy, diverting it away from propelling the bike forward, and likely causing its speed to slow. Too little is dangerous, inviting a spill.

Reactive power occurs only in alternating current (AC) circuitry. Figure 2.6 shows how it is measured. The red and black curves display the sinusoidal time variation of voltage and current, respectively, cycling through 1/60th of a second. The phase angle $\theta$ lag of the current behind the voltage measures the fraction of reactive power in the system. Unlike active power—which we simply call power— reactive power does no work. It is necessary, however, to maintain the magnetic fields cycling at 60 Hz in motors, transformers, transmission lines, and other devices that are based on magnetic induction from alternating current.[7]

A grid's demand from motors, transformers, and much else absorb reactive power. Turbine generators are tuned to generate the needed proportions of reactive power. Too little reactive power causes voltages to fluctuate, flutter, dip and in extreme cases to collapse; too much causes the grid to lose efficiency and electrical

**Fig. 2.6** Current Lag $\theta$ as reactive power measure. (Source: Commons, Wikimedia.org)

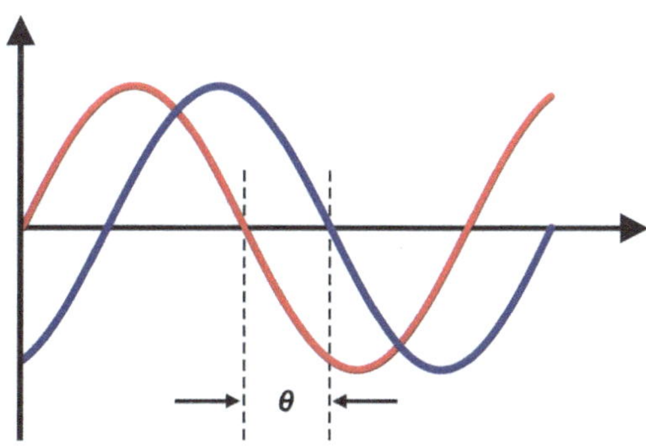


---

[7]For a detailed explanation of reactive power, see for example https://eepower.com/technical-articles/transformer-power-ratings.

equipment to overheat. Either extreme may lead to equipment failure, or worse, blackouts.

Both synchronous inertia and reactive power of turbine generators maintain grid stability. They also support grid reliability when contingencies occur. Beyond the small fluctuations caused by random variation in supply and demand, frequency increase may be caused by a transmission line disconnect resulting in an apparent decrease in demand. Conversely, a frequency drop may be caused by a generator failure resulting in a decrease in supply. The synchronous inertia of large turbine generators slows the resulting frequency deviations from 60 Hz, providing precious time for grid-balancing authorities to take corrective actions, such as rerouting power around a downed high-voltage line or increasing the power of backup generation to replace the failed power plant.

Turbine generators provide another service. They commonly run at somewhat less than full power to provide what is called spinning reserve. Since they are already spinning at 60 Hz, they can increase very quickly to full power. Until slower acting resources can take over the task, grid-balancing authorities depend on spinning reserve to quickly restore frequency stability following major upsets. On grids that include several large turbine generators, each operating with a spinning reserve, it becomes possible to quickly compensate for the loss of the system's largest generating unit. When substantial amounts of wind or solar energy—called variable renewable energy (VRE)—connect to the grid, they often replace fossil fueled turbine generators, causing loss of the stability and reactive power that the turbines provided. That creates challenges to which we shall return.

***

For electric grids to function reliably, in addition to stability, their operators must maintain balance. In other words, supply must always equal demand on a second-by-second basis. That demand varies by the time of day, the week and the season. Balance is maintained by continual automated adjustment of the flow of steam, gas, or water entering each of the turbine generators producing electricity. Assurance of balance is central to many of the characteristics of how balancing authorities operate grids, and to how those operations must change if VRE is to play a major role in combating climate change. The following chapters delve into the nature of these changes. An examination of how grids met the necessity of supply always exactly equaling demand before the advent of VRE provides necessary background.

Figure 2.7 shows the variation in demand—referred to as the load—for a typical week, for three seasons. The load is measured in megawatts of power and time in hours. Fall would be nearly identical to spring and thus is not plotted. Each plot displays 24-h cycles for a week, with daytime peaks when business, government, educational, and other institutions are all using electricity. Minimum demand occurs late at night when there is the least activity, for nearly everyone is sleeping.

These curves come from the warm climate of Texas. Hence, maximum power usage is in summer, driven by air conditioning. Winter is intermediate, with heating and lighting playing leading roles. Note the double daily peaks in winter. These occur because in early morning and late afternoon lighting is needed, even while commerce and industry are functioning. The late afternoon spikes on the spring

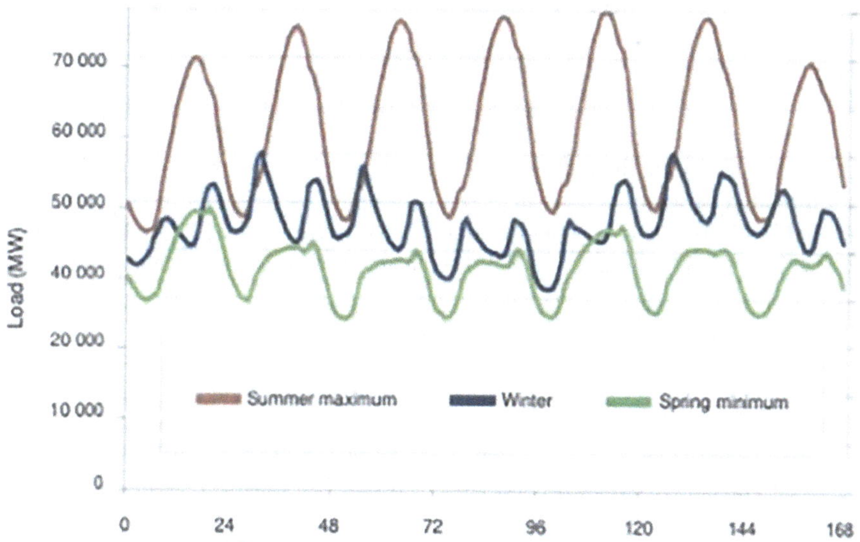



**Fig. 2.7** Variable electrical demand for one week. (Source: NREL)

curve occur as the sun is setting, and street and home lights and appliances are turned on before power use by businesses, schools and other institutions has declined appreciably. In colder regions, winter demand will be higher relative to summer. In New England, for example, peak use may occur on winter nights rather than summer afternoons due to widespread electric heating. Power plants are most often taken offline in spring or fall for maintenance, repair or refueling, when electricity demand for air conditioning and heating is minimal.

The profiles shown in Fig. 2.7 will grow with electricity demand. Their shapes will change in ways difficult to predict with the electrification of motor vehicle transport and other segments of the economy. Likewise, demand may decrease somewhat and its profile change as more families and institutions install solar panels to generate their own electricity, "behind the meter" as it is said to be. Moreover, those who sell the excess electricity from home generation to the local distribution utility will affect both the company's supply and demand curves. While the details of how the time profiles for electricity supply and demand will change may be difficult to predict, they will have little effect on the conclusions to be drawn from the analyses that follow. Our discussion for the remainder of this chapter is limited to electric grids before wind and solar energy became a significant factor in their operation. The following chapters focus on the characteristics of these weather dependent energy sources and their impacts on grid operations.

Before the arrival of variable wind and solar energy, electric grids relied entirely on dispatchable sources of electricity, meaning sources that could be dispatched to meet the demand for electricity. Fossil-fueled, nuclear, hydroelectric and geothermal plants generate dispatchable—also called firm—sources of electricity. Moreover, with the exception hydroelectric power, these sources are all thermal plants: They produce high-temperature heat by fossil fuel combustion, geothermal production, or nuclear fission. A steam or gas turbine then converts the heat to the mechanical energy of a shaft spinning at 60 Hz. A generator then converts the mechanical energy to alternating current electricity.

In thermal plants, temperature differences of hundreds of degrees or more occur across distances of an inch or less. Thermal shock phenomena place limits on how fast these temperatures can be allowed to change. (Think of pouring boiling water into a thin China cup at room temperature, causing it to shatter.) Those limits in turn restrict how fast plants can start, stop and change power levels. Economic and technical tradeoffs intertwine in designing power plants to supply the constantly varying demand for electricity illustrated in Fig. 2.7.

The three classes of plants needed to supply electricity over a typical 24-h demand cycle in the most economic manner are specified in Fig. 2.8: baseload, load following, and peaking plants. (Again, load and demand are the same.) As the relative sizes of the areas in Fig. 2.8 indicate, well over half of the energy delivered is baseload. Baseload plants are nearly always large, sometimes with powers of 1000 MW, that is a gigawatt (GW), or more. They are optimized to run at full power 24 h per day, 7 days per week.

Continuous baseload power operation requires fuel costs to be minimized. Meeting these criteria results in high capital costs, even while their designs sacrifice the ability to change power levels rapidly—they normally take hours to go from zero to full power—and are likely to have a quite large minimum power below which they cannot operate. Nuclear and most coal plants fit into the baseload category as do combined cycle gas-fired plants. In the latter, fuel efficiency is gained by combining gas and steam turbines; the hot exhaust from the gas turbine boils water to fuel a steam turbine.

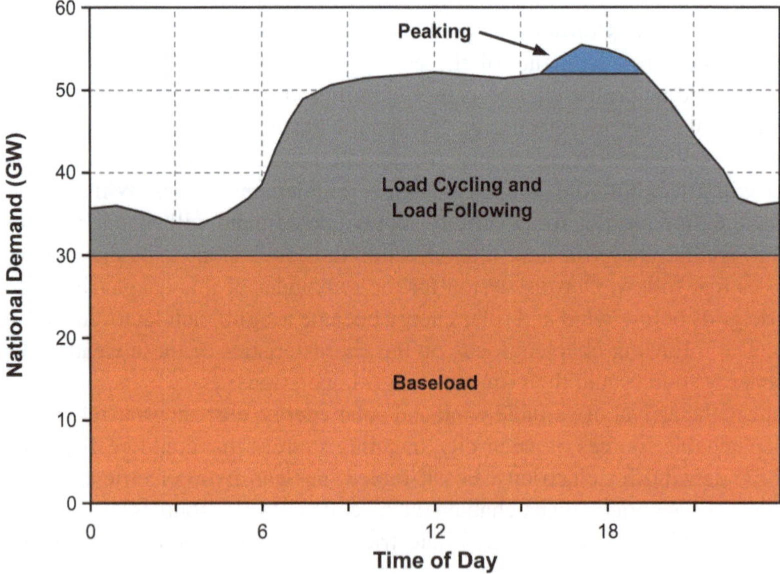


**Fig. 2.8** Variable electric demand for one day. (Source: IAEA)

Plants used for load following have characteristics falling between those of baseload and peaking plants; they are likely to have intermediate capital and fuel costs. Load following plants must be capable of changing power level only fast enough to follow the more gradual of the demand slopes shown in Fig. 2.8. Typically, they must operate continuously even if at very low powers, for once shut down, they may take hours to restart. To a limited extent, nuclear, coal and combined cycle gas-fired plants can be used for load following, although they then are less economical than when providing baseload power. Oil fired plants or dual fuel plants, which can burn gas or oil, may also be used for load following.

Peaking plants fill in the top of Fig. 2.8. They are designed to have relatively low capital cost, but their fuel may be expensive. The higher fuel costs are tolerable since generally they operate only when both baseload and load following plants are already operating at full powers or when demand spikes too rapidly for other plants to keep up with it. Peaking plants must be capable of startup times of only a few minutes, and of the ability to ramp up from zero to full power in a minute or two. Open cycle gas turbines—similar to overgrown jet engines—are the most common peaking plants. Piston engines—overgrown cousins to automobile engines—are called for when even faster ramping times are needed. Peaking plants are less economical to operate than the other plant types. If fossil fueled, they tend to cause more air pollution, and nearly twice as much carbon dioxide per MWh of electricity as plants operating steady-state at near full power.[8]

Not being thermal engines, hydroelectric plants are the most flexible sources of electricity. Their water turbines can ramp up and down in power in minutes, and they produce no carbon dioxide or other air pollutants. Moreover, they can provide baseload, load follow, or peaking power as the grid situation requires.

***

Reliability requires that dispatchable sources have fuel on hand, ready for use. Fossil-fueled plants maintain fuel reserves in the form of large piles of coal, tanks filled with oil, or a pipeline connected to underground natural gas reservoirs. Their fuel inventories must be large enough to supply power during supply interruptions lasting weeks. At nuclear power plants, the bundles of fuel rods that constitute the reactor's core contain many months of supply. Hydroelectric power is dispatchable since months of fuel are on hand in the form of the water confined in a dam's reservoir, and the inexhaustible supply of heat deep beneath the earth's surface fuels geothermal power plants. Run-of-river hydroelectric generators are an exception since they have no reservoirs, making their power totally dependent on the seasonal variations of the river's flow.

Problems can arise in maintaining fossil fuel supplies. Coal piles may freeze over, or rail and barge transport may be disrupted. Weather may also block the transport of oil to on-site tanks. For natural gas, the situation is more complicated since it supplies home heating as well as electricity in cooler climates. That can lead to depletion of storage reservoirs, and more frequently, congestion of pipelines

---

[8] http://euanmearns.com/co2-emissions-variations-in-ccgts-used-to-balance-wind-in-ireland/

delivering gas both for home heating and for producing electricity. To counter the problem, some plants burn either gas or oil. The oil is stored in tanks on site for backup energy. Long-lasting droughts may cause reservoir water levels to drop, severely limiting the power output of hydroelectric plants.

Wind and solar energy, while not fuel dependent, are also subject to weather extremes. Wind turbines can ice over, and at subzero temperatures they automatically turn off to prevent equipment damage and draw power to keep equipment from suffering damage.[9] Snow can obscure solar panels, and hailstorms damage them.[10] Nuclear plants are least susceptible to disruption, for they routinely run for 18–24 months between the replacement of about a third of their fuel. Moreover, the fresh fuel arrives on site weeks if not months before needed to ensure that weather delays do not interfere with refueling.

Beyond assuring that fuel supplies are adequate and maintaining grid stability and balance, those responsible must plan for future power plants and for the transmission lines to get the power they produce to the locations to where it will be consumed. The need for new capacity goes far beyond accounting for population and economic growths. It must account for the closing of fossil fueled plants to reduce to reduce greenhouse gas emissions, the burgeoning needs of data centers to implement artificial intelligence, and the electrification of the transportation and other segments of the economy to meet the 2050 goal of carbon neutrality.

Those responsible for long-term planning must deal with multiple tradeoffs, foremost among them between reliability and affordability. If they build power plants, transmission lines, and more based on estimates of future electricity demand that are too low, capacity will be inadequate, and reliability will deteriorate, increasing the danger of catastrophic blackouts. If their demand estimates turn out to be too high, they are likely to build too many plants and other infrastructure, resulting underutilized facilities. The need for investors to recoup the cost of building such facilities will necessitate consumers' electric bills growing to unaffordable levels.

Whether the cause is grid instability, supply not matching demand, cut off from fuel supply or the failure of long-term planning, the threat of catastrophic blackout looms over the power industry. Blackout's consequences multiply with each hour that elapses before power is restored. Short blackouts lasting only 2 or 3 h are least harmful. For this reason, balancing authorities impose controlled rolling blackouts to avoid the far greater danger of longer lasting uncontrolled loss of electrify.

Even if a blackout lasts only a couple hours, traffic jams, elevator stalls, classroom interruptions, and more will result. But emergency generators assure the surgeries can be completed, and air traffic controllers can land planes safely. Frozen food won't become uneatable, and room temperatures won't become intolerable. But as an uncontrolled blackout's duration lengthens, medical procedures must be postponed, flights canceled, frozen food discarded, and uncontrolled temperatures

---

[9] https://energynews.us/2019/02/27/wind-turbine-shutdowns-during-polar-vortex-stoke-midwest-debate/

[10] https://www.pv-magazine.com/2024/04/17/hail-damage-and-toxicity-risks-in-solar-plants

will aggregate health problems, often leading to fatal results. Pumps, filters and other equipment in water purification and sewage disposal plants will become inoperative, leading to intolerable public health conditions. The extensive losses of power during the weeklong Texas freeze of February 2021 contributed substantially to the hundreds of deaths caused by the storm.[11] Indeed, the developed world is totally dependent on reliable supplies of electricity.

Before recovery from a blackout can begin, its cause must be determined and fixed or circumvented. Only then can the slow and arduous process of recovery proceed; it will last hours if not days. Unfortunately, once nuclear, wind, solar and most fossil fueled power plants have shut down, they cannot simply be restarted. They first require significant amounts of 60 Hz electricity from the grid to power control systems, coolant pumps, and other ancillary equipment. But in a blackout, there is none. Consequently, generators with "black start" capability must be used. Most often these are diesel or gas-fired generators designed to be started by batteries, much like automobiles. Where available, hydroelectric power also has black-start capability. These plants supply 60 Hz power to the grid during a carefully orchestrated sequence of events. Power from back-start plants is used to start larger generators, and neighborhood by neighborhood power is restored, always keeping demand in balance with supply as each of the larger generators resumes operation, most of them taking hours or even days to do so.

---

[11] https://en.wikipedia.org/wiki/2021_Texas_power_crisis

# Chapter 3
# Creating Carrots and Sticks

**Keywords** Subsidies · Tax credits · Portfolio standards · Contracts for difference · Flat tax · Cap-and-trade · Electricity markets · Long-term investment · Variable renewable energy · Dispatchable energy · Deregulation · Nuclear energy

"Renewables are on track to satiate the world's appetite for electricity" reports the Washington Post. "New York takes a big step toward renewable energy in 'historic' climate win," headlines the Guardian. "The White House set out a target of 80% renewable energy generation by 2030," broadcasts NPR. "You can help build a better, cleaner world for all of us by switching to 100% renewable energy," touts Clean Choice Energy's solicitations for investments in wind and solar farms. The Green Power Markets for Renewable Energy Certificates (RECs) are explained on an Environmental Protection Agency webpage. In the public's mind, a single word, renewables, has been synonymous with the fight to stem climate change and has frequently been used together with clean, green, sustainable, and other laudatory adjectives.

In the media and among the public, enthusiasm for renewables is focused on the plummeting cost and growing presence of wind turbines and solar photovoltaic panels across our landscapes, such as pictured in Fig. 3.1. The US Department of Energy refers to wind and solar energy as variable renewable energy (VRE) to indicate their intermittency and hour-by-hour, day-to-day, and seasonal weather dependence. Hydroelectric and geothermal energy are also renewable. In contrast to VRE, however, hydroelectric and geothermal power, along with nuclear, fossil, and biofueled energy are designated as dispatchable or firm energy because they can be dispatched on demand, regardless of day-to-day weather.

While VRE is at the center of renewables enthusiasm, hydroelectric power worldwide is the largest provider of electricity that is free of greenhouse gases.[1] In Europe, nuclear reactors supply 25% of the electricity and 50% of the low-carbon

---

[1] https://www.iaea.org/bulletin/what-is-the-clean-energy-transition-and-how-does-nuclear-power-fit-in

**Fig. 3.1** Wind turbines and solar panels. (Source: iStock)

electricity.[2] In the USA, nuclear is the largest low-carbon energy source, supplying 18.9% of the electricity, followed by wind and solar energy at 13.8%, and hydro-electricity at 6.3%, respectively.[3] Several large nations have ambitious goals on the fraction of their electric power that will be renewable by 2030: China 28%, European Union 42.5%, India 40%, UK 60%, while in the USA, targets vary state by state both in percentage and target date. Taken together, these nations account for 60% of the world's electricity consumption.[4]

How nations choose to decarbonize their electric grids depends a great deal on the natural resources they have available. Norway, for example, obtains nearly all its power from hydroelectric plants, Sweden and Canada from hydroelectric and nuclear power, and Iceland from geothermal energy. For many nations, however, the paucity of suitable land to expand hydroelectric or geothermal power limits their generation to satisfying only small fractions of the demand for electricity. Hence, wind and solar energy are the renewables that must partner with nuclear energy in replacing fossil fuels and meeting the growing demand for electricity.

Complex interactions of technical, economic, social, and political issues will determine what the fractions of renewables and nuclear will be, and how the percentages are likely to change in future years as fossil fueled plants are retired and carbon-free generation becomes dominant. As a baseline for the coming decades, it

---

[2] https://www.euronews.com/business/2024/05/16/economics-of-nuclear-power-the-france-germany-divide-explained

[3] https://www.nei.org/advantages/climate

[4] https://www.visualcapitalist.com/progress-on-2030-renewable-energy-targets-by-country/

is instructive to compare the present states of two somewhat similar nations, but one utilizing only renewables, while the other relies predominantly on nuclear energy.

Germany's policy has been to rely solely on renewables. Consequently, the last of the nuclear reactors that once provided 20% of its electricity was shut down recently. Heavy investments in wind and solar energy have led to almost half of its electricity being produced by VRE, with most of the remaining coming from fossil fuels. The German goal is to have 100% of its electricity coming from renewables by 2035. In the 1980s, France transformed its electric grid to be powered primarily by nuclear energy, thus freeing itself from unreliable foreign oil suppliers. And indeed, as much as 75% of its electricity has come from nuclear reactors, and the planning for future reactors is ongoing. A substantial part of France's remaining electricity comes from hydroelectric plants, along with wind and solar generation.

What have the results been to date? The price consumers pay for electricity in Germany is nearly double what it is in France, and worse, nearly ten times as much $CO_2$ per kilowatt hour is released to the atmosphere from electricity produced in Germany than from that generated in France.[5] Moreover, France has been a net exporter of electricity, while Germany operates once shuttered coal plants to compensate for the power deficit caused by shutting down its nuclear reactors.

***

No matter whether renewables or nuclear will plays the larger role in years to come, governmental policy initiatives are needed to accelerate the decarbonation of electric grids. Such policies generally fall into the categories of either carrots or sticks. The carrots certainly include supporting basic research and development through demonstration projects at the pre-commercialization level. They are essential if full advantage is to be taken of potentially game-changing technologies that presently lie over the horizon. Nuclear fusion and deep geothermal energy are likely examples. Beyond research and development, subsidies are needed to support carbon-free technologies until they can compete with those that release greenhouse gases to the atmosphere.

In the USA, at both federal and state level, subsidies are more successful than penalties in navigating legislative processes, in part because some types of subsidies camouflage to some extent who is paying for them. At the federal level, the most direct subsidies are in the form of investment and production tax credits. These allow companies to reduce their taxes in proportion to their capital and operating costs. They amount to 30% of construction costs and 26 dollars per megawatt hour (MWh) of VRE produced. Accordingly, they are financed by federal taxpayers. Since they are buried in the thousands of pages of the federal budget, they tend not to attract a great deal of taxpayer or legislative opposition.

At the state level, subsidies most commonly take the form of portfolio standards. These require electric distribution utilities to demonstrate that at least a specified percentage of their power comes from carbon-free sources. These range between a few percent and 50% or more, and often the percentage is increased incrementally

---

[5] https://environmentalprogress.org/big-news/2017/2/11/e

on a year-to-year basis. The system works as follows. The companies generating electricity from wind or solar energy issue a Renewable Energy Credit (REC) for each MWh of energy that they generate. RECs are sold on a market to the distribution utilities, who must prove they are meeting the portfolio standard for the states in which they operate. These costs are then passed on to their customers, appearing as an item—along with production, transmission, and distribution—on monthly electric bills. As such, they do not impact state government budgets and hence encounter relatively minor legislative resistance.

While the carrots of subsidies may assist decarbonized energy in competing with fossil fuels in the electricity marketplace, they aren't likely to suffice in reducing fossil fuel combustion at the rate needed to meet the Paris Accords. Sticks will almost certainly be required. The simplest stick would be to impose a tax-based price on emissions, such as dollars per ton of $CO_2$ ($/t$CO_2$) released to the atmosphere. The current price estimate used by the US government is 51 $/t$CO_2$ However, a comprehensive review of science studies concluded that it should be 3.6 times higher at 186 $/t$CO_2$.[6] Such estimates are based not only on power plant emissions; they also include transport and other factors. The US Congress, however, seems to have no appetite for legislating such a tax.

Rather than imposing a flat tax for each ton of $CO_2$ emitted, various cap-and-trade regimens have been adopted in the EU, UK, China, New Zeeland, South Korea, and other countries. While Congress hasn't taken action to implement a national program, a growing number of states are adopting cap-and-trade programs through the Regional Greenhouse Gas Initiative (RGGI).[7] With state authorization, RGGI auctions emissions allowances—sometimes called credits—each year. These must be obtained by all fossil fueled power plants with installed capacities of more than 25 MW for each ton of $CO_2$ emitted annually. Trading of allowances provides plant owners flexibility in meeting the requirements. If it is too expensive for a plant to meet the allowance requirement—such as an old plant scheduled for retirement within a few years—the owner can buy allowances from the owner of a plant that has emitted less than its allowance permits. Hence, owners of less polluting plants are rewarded, and those of more polluting plants are penalized. Each year the emission allowances are reduced. Thus, RGGI regulated plants reduced emissions by nearly half over a 10-year period, and the states in which they operate have set an emissions reduction goal of 30% between 2020 and 2030.[8]

Federal and state policies reflect the opinions of their constituents. Until quite recently, the widespread enthusiasm for renewables led to heavy emphasis on VRE subsidies to ensure that wind and solar energy could gain a hold on electric grids where fossil fueled plants were more economical. Nuclear, however, had no such advantage, for until recently much of the public has either opposed, ignored or given

---

[6]Rennet, K., et al. (2022) Comprehensive evidence implies a higher social cost of $CO_2$. *Nature* 610, 687.

[7]https://www.c2es.org/content/cap-and-trade-basics/

[8]https://www.c2es.org/content/regional-greenhouse-gas-initiative-rggi/

acceptance only grudgingly, assuming nuclear may be necessary even if unwanted. Vehement opposition came from the Sahara Club, Friends of the Earth, Greenpeace and other environmental advocacy groups. Such organizations stoked public apprehensions of radiation related risks and used the courts to delay or defeat nuclear initiatives. For example, court mandated changes in licensing requirements, while plants were being built often led to schedule slip and cost overruns.

Only VRE was allowed to take advantage of federal tax credits. At the state level, portfolio standards mandated growing fractions of electricity that must come from renewable sources were applicable only to wind and solar energy. One estimate is that subsidies for wind and solar energy are nearly 30 times higher than they are for nuclear energy.[9] Competing against both historically low natural gas prices and subsidized renewables, nuclear plants became uneconomical to operate, and their owners' threatened closures. To avoid losing major sources of carbon-free electricity, the Illinois legislature created carbon mitigation credits to keep some nuclear plants running. Similar programs followed suit in Connecticut, New Jersey, New York, and Ohio, all states relying heavily on nuclear power plants for carbon-free electricity.

Across the world, as in the USA, efforts are underway to provide subsidies that will allow carbon-free electricity generation to compete with fossil fuels. The 28 nations of the European Union (EU) play a prominent role in these efforts. Central to the EU system are auctions of grants to subsidize renewables. The financing mechanism pools contributions from contributing countries to build and operate facilities in host countries for carbon-free electricity generation, transmission, transportation, storage, heating, and more. The contributing and host country may be the same, but cross-border agreements are also encouraged. In the latter case, some fraction of the investment is awarded to the contributing country's goal of achieving 42.5% renewables by 2030. Thus far, nuclear energy has not been included in the subsidy program because of disagreements between nations, principally France and Germany, as to whether nuclear energy should be considered "green" under EU rules.

Other nations have varying methods for encouraging development of low-carbon electricity generation. We take the UK and Australia as examples. In the UK, a government authority auctions long-term, typically 15 year, contracts to renewable and nuclear developers.[10] The contracts contain a strike price, which is substantially higher than the fluctuating market price of electricity. The contract pays the developer the difference between the strike price and the market price. If the market price should rise above the strike price, however, the developer must pay the government authority difference between the two. In this way, the developer is protected from market prices that are too low to generate a reasonable profit. Likewise, if market prices become excessive, the developer must return some of the profit to the

---

[9] https://www.instituteforenergyresearch.org/fossil-fuels/renewable-energy-still-dominates-energy-subsidies-in-fy-2022/

[10] https://www.empireengineering.co.uk/contracts-for-difference-what-are-they-and-how-do-they-work/

governmental agency. Understandably, these arrangements are called contracts for difference.

Most subsidies in Australia have some similarities to those in the UK. The governing body auctions agreements with developers with a revenue floor and ceiling.[11] If the market price is below the floor, the developer is paid a specified fraction of the difference. Conversely, if revenue exceeds the ceiling, the developer pays the governing body an agreed upon fraction of the difference. Both UK and Australian systems have the benefit of providing some measure of security in making long-term investments in large electricity decarbonization projects.

***

Traditionally, within a specified geographical region, electricity was generated, transmitted and distributed by a single monopoly called a vertically integrated utility. The rates it could charge for electricity were regulated by a governmental commission. While such arrangements provided electricity reliably, with few blackouts, consumers often thought electricity prices were too high. Consequently, during the decades of restructuring and deregulation of communications, air travel, gas and other industries, the delivery of electricity was restructured and deregulated in large parts of the USA as well as in many other nations.

Deregulation encompasses seven of the ten US grids, labeled in Fig. 2.2 as CAISO, ERCO, ISO-NE, MISO, NYISO, PJM, and SSP. In each, a nonprofit-balancing authority operates a wholesale electricity market that adheres to rules of the Federal Energy Regulatory Commission (FERC).[12] The remaining three grids, NORTHWEST, SOUTHEAST, and SOUTHWEST, remain regulated by state utility boards or similar governmental organizations. In these areas, dozens of utilities, many of them vertically integrated, supply electricity to designated areas.

In deregulated areas, vertically integrated utilities were required to sell their power plants to companies that they didn't control. The high-voltage transmission networks remained privately owned monopolies, operated and controlled by non-profit grid authorities under FERC rules. A grid authority is often called a balancing authority since it is responsible for guaranteeing that electricity supply is always exactly equal to demand. Its related function is of operate a wholesale electricity market. The owners of power plants, now called merchant generators, bid to sell their electricity on these competitive wholesale markets. Distribution utilities—sometimes what remained of the vertically integrated utilities—then buy the electricity at market price and distribute it to industrial, commercial and residential customers.

The net result of deregulation has tended to be electricity becoming less expensive on average, but with more volatile pricing, and a supply that may be less reliable. Sorely absent from the rules under which FERC electric markets operate are

---

[11] https://www.dcceew.gov.au/energy/renewable/capacity-investment-scheme/information-for-proponents

[12] https://www.ferc.gov/introductory-guide-electricity-markets-regulated-federal-energy-regulatory-commission

effective mechanisms to encourage long term investments in new power plants. These are needed to assure adequate capacity to meet future power needs. The absence of financial incentives has led to a phenomenon called "the missing money problem," a topic to which we shall return in Chap. 16.

***

With the dangers of climate change becoming all too apparent, and the need for nuclear energy to assume a major role in rapidly reducing greenhouse gas emissions, attitudes worldwide and in the USA are changing. More than 30 nations, including the USA, have endorsed a declaration to triple the world's nuclear energy capacity by 2050.[13] In public parlance, "clean "energy is used increasingly instead of "renewable" energy to acknowledge the importance of nuclear power, and government policies are treating renewables and nuclear in a more evenhanded manner. The US Inflation Reduction Act provided a nearly 400-billion-dollar stimulus for curbing climate change. By making subsidy programs technology neutral, government programs have placed renewable and nuclear projects on a more equal footing. Now, for example, federal investment and production tax credits are available to both. Debate will continue on which should play the larger—or even predominant— role in fighting climate change. Technical, economic, and social issues can now be addressed on a more level playing field in determining how renewables and nuclear should be paired in the fight to curb climate change.

---

[13] https://world-nuclear.org/news-and-media/press-statements/six-more-countries-endorse-the-declaration-to-triple-nuclear-energy-by-2050-at-cop29

# Chapter 4
# Energy from the Sky

**Keywords** Wind energy · Solar energy · Photovoltaic cells · Efficiency limits · Variable renewable energy · Dispatchable energy · Rooftop energy · Direct and alternating current · Maintenance · Blackout · Grid stability

The wind turbine's American debut was spectacular. With its picture on the cover, the December 20, 1890, *Scientific American* hailed it "As an example of throughgoing engineering work it cannot be excelled" and "With the exception of the gigantic windmill and electric plant shown in our engraving, we do not know of a successful system of electric lighting operated by means of wind power." Only a year after Thomas Edison invented the incandescent light bulb, Charles F. Brush built the massive structure shown in Fig. 4.1 in Cleveland, Ohio.

The turbine was far larger than the many windmills across North America, widely used for pumping water. Its tower was 60 feet tall, and its 144 blades made up a circle with a diameter of 56 feet. Thus, the structure reached 88 feet into the sky—nearly the height of a nine-story building. Inside the tower, a shaft turned a system of pulleys and belts that spun a dynamo (i.e., an electric generator) at 500 cycles per second, converting mechanical energy to direct current electricity stored in more than 400 batteries in Brush's basement. The apparatus could transmit as much as 12 kilowatts (kW) of power to incandescent or arc lights, and to electric motors.[1] Brush also invented an improved dynamo for converting mechanical to electrical energy and founded the Brush Electric company, which in 1892 merged with Edison's General Electric Company to form the company named General Electric.

The Cleveland behemoth was not the world's first wind turbine. That honor goes to James Blyth of Glasgow, Scotland, who in 1887 built a 30-foot-tall cloth-sailed turbine on the grounds of his holiday cottage to power its arc lighting. In Europe, a succession of advances followed Brush's work. Poul La Cour, a Danish scientist, greatly improved wind turbine efficiency in 1899. He found that a small number of relatively narrow rotor blades were far more efficient in converting wind energy to

---

[1] https://www.cleveland.com/metro/2011/08/charles_brush_used_wind_power.html



E. E. Lewis, *Renewables or Nuclear*, https://doi.org/10.1007/978-3-032-08074-5_4

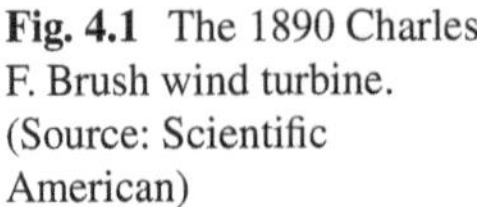

**Fig. 4.1** The 1890 Charles F. Brush wind turbine. (Source: Scientific American)

electricity. It is generally agreed that his design was the first in a line of modern turbines for generating electricity. The work of La Cour and others established Denmark as a leading center for wind turbine research and development. In 1919, the German physicist Albert Betz placed turbine design on a sound theoretical basis. He proved what became known as the Betz's Law.[2] It states no more than 59% of the wind energy impacting the circular area formed by the turbine's rotating blades can be converted to mechanical energy. Wind turbines in the kw range spread over Denmark, and export would follow later.

Patterned after the traditional American farm windmill, wind turbines spread across the country in the 1920s and remained for pumping water and providing intermittent electricity until the more remote rural areas were connected to grid electricity decades later. Utility scale turbines got a boost in 1973 when the middle eastern oil boycott and subsequent hikes in fuel prices increased interest in wind as an alternative form of energy. However, interest faded as oil and other energy sources became more economical. But then in the late 1970s when oil and gas scarcities developed, serious consideration of wind energy came to the forefront.

Beginning in the early 1980s, the first utility-sized wind farm in the USA was erected in Altamont Pass, California. Not surprisingly, highly regarded Danish turbines, known for reliability and efficiency, were the first to occupy the farm. The farm grew to more than a thousand turbines of various sizes and manufacture. Most were small by today's standards but together reached an installed capacity of nearly 600 megawatts (MW).[3] Federal tax credits and other subsidies combined with steep

---

[2] https://www.alternative-energy-tutorials.com/wind-energy/betz-limit.html

[3] https://www.acgov.org/cda/planning/landuseprojects/windturbineproject.htm

reduction in turbine capital cost per MW generated led to the widespread increase in wind farm capacity across the nation.

***

Modern wind turbines have three blades, as diagramed in Fig. 4.2. They continue to grow in size, since their power is proportional to the circular area swept out by their blades. Those for offshore locations are the largest, approaching installed capacities expected eventually to reach 20 MW. The distance from the ground to the center of the turbine's rotor may exceed the length of a football field, with a rotor diameter measuring nearly 50% more than the rotor height, about as tall as the Great Pyramid of Giza.[4] Onshore or off, the turbines are a marvel of modern technology, their design spanning nearly every engineering discipline.

Each blade of a modern wind turbine has a cross section similar to that of an airplane wing, gaining lift from its shape to propel its rotational motion perpendicular to the wind direction. In fact, the blade's shape is even more complex than a wing, for a gradual twist in its cross-sectional shape takes place between its base of the rotor and its tip. The twist is necessary, for while the blade's rotation is constant, its speed increases from a minimum at the rotor to a maximum at its tip. Without the twist, the turbine would be much less effective. In fact, in some ways, wind turbine blades are more akin to those of an aircraft propeller than a wing because they both

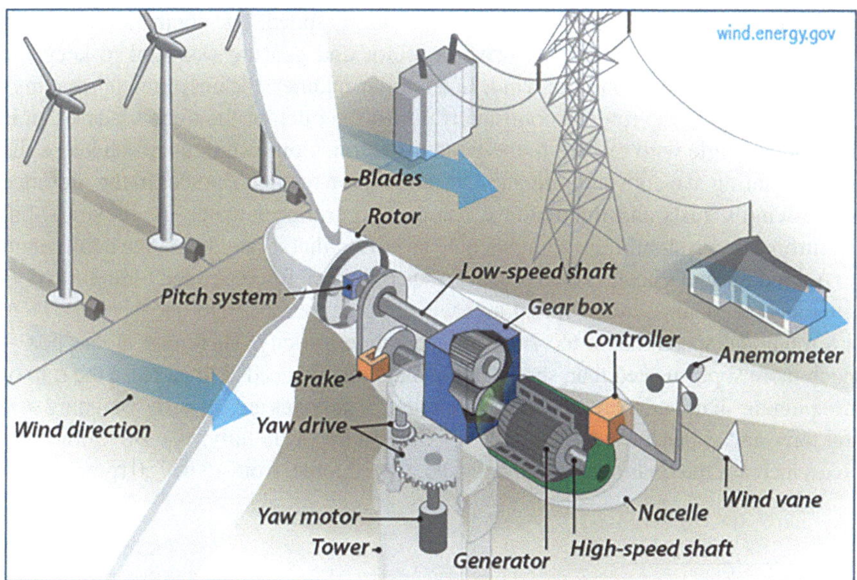


Fig. 4.2  Wind turbine nacelle. (Source: DoE)

---

[4] https://www.eia.gov/energyexplained/nuclear/us-nuclear-industry.php

rotate on a plane perpendicular to the air flow direction. The difference, of course, is that on an airplane the engine rotates the propeller to drive the aircraft forward. The blades of a wind turbine capture the energy of the oncoming wind to drive an electric generator.

The blades rotate at about 20 revolutions per minute (rpm). They start turning with a wind speed of about 8 miles per hour (mph). At greater wind speeds, they rotate faster, with the power produced multiplied by a factor of eight for each doubling of wind speed until the turbine reaches its rated power at a wind speed of about 30 mph. For wind speeds greater than 30 mph, the turbine's rotational rate is held constant to prevent increasing speed from causing excessive blade stresses. Operating at constant power, the turbine loses efficiency as wind speed increases.[5] However, the decreasing efficiency isn't important because the fuel is free; it's the wind. Moreover, even across the windy Great Plaines, the wind speeds averaged over a year are less than about 20 mph,[6] substantially less than the 30 mph required for wind turbines to reach full power.[7] Turbines must shut down to avoid equipment damage if wind speed surpasses about 60 mph. In principle, turbines could operate at higher wind speeds. However, the wind blows at higher speeds for too few hours each year to recover the costs of the stronger blades that would be required.

The nacelle, about the size of a large garage, sits atop a turbine's tower. It is the nerve center of turbine control. The simplified diagram shown as Fig. 4.2 illustrates the larger pieces of mechanical, hydraulic, and electrical equipment, packed in the nacelle.[8] They are tied together and controlled by on-board computers. The wind vane detects wind direction, and the anemometer its speed. The controller uses wind vain readings to power the yaw drive, a motor and gearing essential to keep the turbine facing directly into the wind. In a similar manner, the anemometer readings cause a system in the rotor to continually adjust the pitch of the three blades, that is the blades' angle with respect to the wind direction. Controlling the pitch keeps the blades rotating at a constant rate, thus stabilizing the turbine's power in the presence of the wind's gusts and fluctuations. Another controller actuates the brake to shut the turbine down should the wind become so strong that it could damage equipment.

A wind turbine's power is the product of its rotational force—called torque—and its rotational speed, measured in rpm. Large wind turbines produce a torque that is too large at a rotational speed that is too small to generate electricity. Therefore, a gear box—much more sophisticated than shown in the diagram—is needed to increase the rpm of the spinning shaft and thereby reduce its torque. Gear boxes are massive, weighing several tons, and expensive, amounting to 20–30% of the wind turbines cost.[9] Some gear boxes increase the rotational spin by 50-fold, for example, from 15 to 750 rpm.

---

[5] For graphs of power and rotor speed see https://www.nrel.gov/docs/fy12osti/52409-2.pdf.

[6] https://www.landgate.com/news/average-wind-speeds-by-state

[7] Ibid.

[8] https://energy.gov/eere/wind/how-do-wind-turbines-work

[9] Ancona, D. and McVeigh, J. (2001). Wind Turbine - Materials and Manufacturing Fact Sheet. Prepared for the Office of Industrial Technologies, U.S. Department of Energy.

At this higher speed, the spinning shaft connects to the generator, diagramed in Fig. 4.2, where its energy is converted to alternating current (AC) electricity. But even at 750 rpm, the shaft produces only 12.5 Hz AC, still much lower than the 60 Hz needed for the grid. Moreover, even if the generator could produce 60 Hz AC, wind gusts and fluctuations would make the frequency too unstable to put on the grid. Consequently, before the electricity can be fed onto the grid, it must pass through a rectifier that converts the AC to direct current (DC). The DC must then be converted back to AC and synchronized to 60 Hz before connection to the grid.

Wind turbines producing a MW or more of electricity tower into the sky. As a result, performance of periodic maintenance every 6 months to a year within the nacelle and on the turbine blades, as shown in Fig. 4.3, is challenging with technicians performing their duties hundreds of feet above ground level. The turbines are designed to last 25–30 years. The gear boxes, however, degrade rapidly, needing replacement or major overhaul in about 10 years or less.[10] Replacement or overhaul is a major undertaking, having to be performed in a nacelle atop a tower that may be more than 500 feet tall.

Because gear box replacements are more difficult at sea on offshore turbines, much effort is being directed toward eliminating the gear box in what is called a direct drive turbine.[11] The generator in the nacelle must be replaced with one that can operation at much higher torques and a much lower frequency. Current efforts are focused on designs that will meet these criteria with the use of large numbers of paired permanent magnets.[12] At present, it appears that direct drive generators will

**Fig. 4.3** Maintenance of large wind turbine blades. (Source: iStock)

---

[10] Jantara Jr., V. L., et al. (2020). "A Damage Mechanics Approach for Lifetime Estimation of Wind Turbine Gearbox Materials," International Journal of Fatigue **137**, 105671.

[11] https://www.energy.gov/eere/articles/advanced-wind-turbine-drivetrain-trends-and-opportunities

[12] https://www.academia.edu/26665327/Possible_solutions_to_overcome_drawbacks_of_direct_drive_generator_for_large_wind_turbines

be much larger than ones that they replace and weigh more than the present generators and gear boxes combined.[13]

$$* * *$$

In 1839, Edmond Becquerel discovered the first photo voltaic effect to produce electricity directly from sunlight.[14] He coated platinum electrodes with silver chloride and placed them in an acidic solution. Shining a light on the solution caused a voltage between the electrodes and a current to flow through the wire connecting them. In 1873, Willoughby Smith demonstrated selenium's light sensitivity by observing that when exposed to sunlight its conductivity increased. In 1883, Charles Fritts, an American inventor, created the first working photovoltaic cell. Its composition of selenium sandwiched between an iron plate and a semitransparent gold coating, however, resulted in an efficiency of less than 1%. Fritts also created the first solar panel by connecting several of the selenium modules in an array on the rooftop of his New York apartment.

A hiatus in the development of photo voltaic cells, which are now called solar PV (or in these pages, simply solar cells), lasted until 1954. In that year, three scientists working with silicon transistors recently discovered at Bell Laboratories made a solar cell having 6% efficiency, high enough to run electrical equipment. A boost came to solar PV development when NASA expanded research and in 1958 launched Vanguard I, the world's first solar powered satellite. It contained 6 solar cells that powered one of the two transmitters. NASA continued its research and development, deploying large solar arrays on satellites. Their high power to weight ratio made them ideal for aerospace applications. The cells found a wider variety of applications in navigation and telecommunication devices. However, for utility scale power generation on electric grids, weight was not a problem, but their low efficiency and high cost were. There was still much work to be done.

In the decades that followed, prices of solar cells came down, paralleling those of integrated circuits, and expanded research by the Department of Energy and other agencies led to substantial increases in efficiency as well. The antecedents of today's solar farms were built in California. In 1982, the first 1 MW solar PV park was located near Hesperia, followed in 1984 by the 4.2 MW farm on the Carrizo Plain. Today's solar panels have efficiencies of 15–20%, which is getting closer to 33.7%, which is the theoretical maximum efficiency set by the Shockley–Queisser limit for the silicon p–n junction technology on which they are based.[15] Research into alternatives to p–n junction technology is ongoing, leading to laboratory experiments in the low 40% efficiency range. However, these still face serious cost, durability and other issues.

In a solar farm, thin solar cells, each only a few inches in length and width, and producing little more than a volt of electricity, are assembled in arrays and then prefabricated in panels. Each panel may contain a hundred or more cells. Within a panel, the cells are usually wired in series so that their voltages add, although the

---

[13] Ibid.

[14] https://www.sciencedirect.com/topics/engineering/photovoltaic-effect

[15] https://ph.qmul.ac.uk/sites/default/files/u75/Solar%20cells_environmental%20impact.pdf

panels may be wired together in series or parallel respectively to increase voltage or current as needed. A thin layer of nonreflecting glass normally covers the cells to maximize their efficiency.

To understand how the cells work, we must enter the realm of semiconductor physics. For purposes of illustration, Fig. 4.4 shows a small panel consisting of only 21 cells along with the diagrams of a cell's inner working. If readers are satisfied in just knowing that the cells work, they are forgiven if they choose to skip the following three paragraphs.

As the figure indicates, a solar cell contains adjoined layers of two types of silicon semiconductors. The n-type contains added atoms, such as phosphorus, that have one more electron than silicon. Chemical bonding does not constrain those extra electrons, allowing them to roam free within the silicon. In the p-type semiconductor, the added atoms, such as gallium, have one less electron than silicon. The absence of an electron, relative to silicon, appears as a vacancy, which is called a hole.

Within the small distance on either side of the interface between the two types of semiconductors, called the depletion zone, excess electrons from the n-type semiconductor cross the interface and fill the vacancies in the p-type semiconductor. The extra electrons filling the vacancies of the p-type semiconductor form negative ions with their partnered gallium atoms, and these carry a negative charge. The loss of the electrons from the n-type conductor causes it to contain phosphorous ions, which are positively charged. The opposite charges on either side of the depletion zone create an electric field that prevents electrons farther from the interface from escaping from the n-type to the p-type semiconductor.

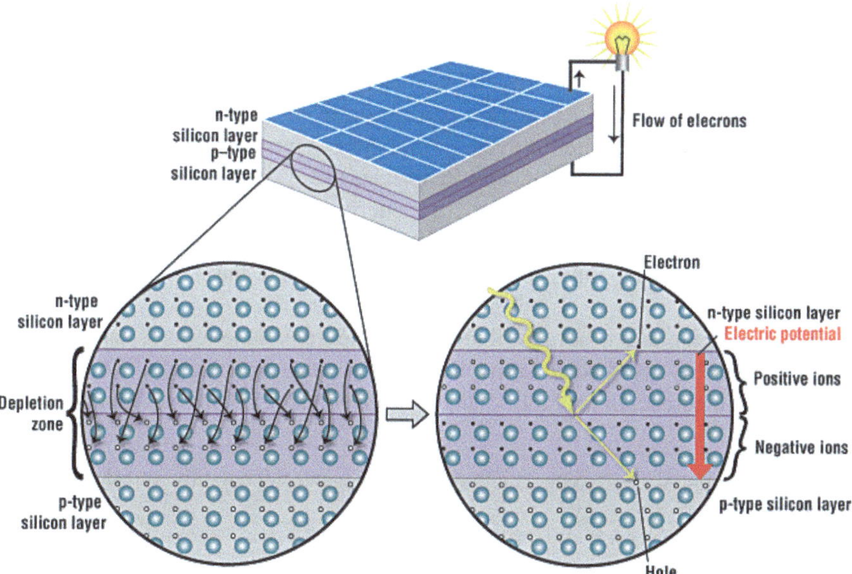


**Fig. 4.4**  Diagram of a solar panel. (Source: American Chemical Society)

The energy of photons of sunlight striking the solar cell causes electrons to be ejected from the silicon. Consequently, the escaping electrons leave vacancies behind. If the vacancies occur in the depletion zone where the electric field is present, the field will move electrons to the n-type layer and holes to the p-type layer. If circuitry is hooked to the metallic connectors at the top and bottom of the cell, electrons will cross the depletion zone from the n-type layer to the p-type layer and then flow through the external circuitry back of the n-type layer, completing the circuit.

Solar farms typically contain row upon row of panels to obtain a specified voltage. The power generated will then grow in proportion to the total number of panels. The panels produce direct current power which must be converted to alternate current synchronized to 60 Hz before connection to a grid. Typically, solar farms are expected to operate for 25–30 years. However, the forming of microcracks causes the panels' efficiency to deteriorate with time, but usually at no more than about 1% per year. Thus, even while the panels' efficiency may deteriorate by 30% in 30 years, it may be economical to keep them in service for a longer stretch of time.[16]

In contrast to employing the solar photovoltaic cells, also called PV cells, which we have been discussing, electricity may be produced from the sun's energy in concentrated solar energy plants. In these, large arrays of mirrors focus on a point or line where solar energy heats a circulating fluid. The hot fluid then drives a turbine generator that produces electricity. In recent decades, solar PV has far outdistanced concentrated solar in the production of electricity. Hence, for brevity, in the chapters that follow, we will refer to solar PV energy simply as solar energy. Chapter 8 includes a closer examination of concentrated solar energy.

***

The smallest wind turbine capable of generating enough power to supply a single home would require an acre or more of land and reach far above the rooftop.[17] Thus, small turbines generating less than a percent or so of a MW are not suitable for urban areas, and small wind turbines in rural areas generate only a tiny fraction of the hundreds of MW of installed capacity of large utility-sized wind farms. The situation is quite different, however, for solar energy. While utility scale solar farms occupy the largest market share at 57%, smaller scale installations owned by commercial and industrial interests account for 23%, and the residential segment is 19%.[18]

While roof top and other distributed solar facilities generate substantially less power than utility-sized solar farms, they are growing more rapidly.[19] Along with the increasing number or roof top solar panels on single family homes, opportunities

---

[16] https://www.paradisesolarenergy.com/blog/solar-panel-degradation-and-the-lifespan-of-solar-panels

[17] https://windexchange.energy.gov/small-wind-guidebook

[18] https://www.iea.org/energy-system/renewables/solar-pv

[19] https://www.gresb.com/nl-en/community-solar-is-the-fastest-growing-segment-of-the-us-solar-market

exist for much larger installations on the flat roofs of big box stores, and on warehouses and other industrial buildings. Adding roofs covered by solar panels over large parking lots would both generate electricity and protect the vehicles from the sun; perhaps with battery storage, nearly self-contained charging stations could be facilitated.

The relationship between distributed solar energy generation and the grid is quite different from connecting utility scale wind or solar farms to high-voltage transmission lines. With batteries installed along with automated grid monitoring apparatus, homeowners can store energy during periods when the price is low and use it when the price is high. In the deregulated electricity markets, such as that exist in much of the USA and several other countries, distributed solar generation would appear as a decrease in the distribution utility's demand rather than an increased supply from the wholesale electricity market.

The situation becomes more complicated when distribution utilities must buy excess electricity from roof-top generation. This can cause technical problems that are difficult to overcome, for until recently transformers, switching circuitry and more were designed for the one-way distribution of electricity from the high-voltage grid to the consumer. But with rooftop generation, the distribution utility must deal with power flows that are reversed through substantial parts of its circuitry. Moreover, in the current practice of net metering, the utility most often must pay consumers the same retail price for the electricity from rooftop generation that consumers pay to the utility for electricity. Consequently, if homeowners sell enough electricity to pay for what they are buying, they owe the utility nothing, even though they are using its lines, transformers, meters and other equipment to buy and sell electricity.

Utilities object to such net metering practice, for not only does it decrease their sales and allows free use of their distribution network, it also disproportionately penalizes those whose lack of wealth or living accommodations prevents them from installing home solar systems to take advantage of the net metering practice.[20] One possible solution would be to separate the utility's charges into a fixed monthly fee for building and maintaining the distribution lines, transformers and other services and equipment along with a second charge per net kilowatt hour of electricity sold to the consumer. Future smart grid technology, such as the virtual power plants (VPPs) discussed in Chap. 9, may ameliorate or eliminate the forgoing conundrums.

As grid structures are modernized and electric vehicles become a larger share of the automotive market, car batteries may have two-way interactions with distribution utilities. Where solar energy provides a substantial fraction of power generation, parked vehicles' batteries may charge during midday hours, while an overabundance of solar energy is being generated, and discharged during nighttime hours. Such an arrangement would serve to mitigate the day-night demand dichotomy that arises with large shares of solar energy. However, overcast skies and

---

[20] https://www.nytimes.com/2017/07/08/climate/rooftop-solar-panels-tax-credits-utility-companies-lobbying.html

unforeseen needs to make long car trips could limit the fraction of battery charge that could be safely allocated to such a regimen.

***

Wind turbines and photovoltaic solar panels are maturing technologies in that fundamental laws of physics limit further improvements in efficiency. Betz's law limits wind turbine efficiency to 59.3% of the wind in the area swept out by the turbine blades, no matter what form wind turbines and their blades take. At present utility scale, blades extract as much as 85% of that limit.[21]

Photovoltaic solar panels are limited by fundamental thermodynamics to converting less than 86.8% of the sun's intensity to electricity. But there is potential for improvement. Perovskite materials, for example, can capture high-energy blue light more efficiently than silicon.[22] Current panels operate at between 20 and 30% efficiency, while at the National Renewable Energy Research Laboratory, 40% efficient panels have been built and tested.[23]

Beyond increasing efficiencies, other advances may lead to further reductions in cost, reduced susceptibility to adverse weather conditions, and other improvements in performance, endurance, and reliability. Although much of the technology is already highly optimized, examples abound: For wind turbines, advances may include eliminating the need for gear boxes or the necessity of first generating DC power and then converting back to AC before grid connections, reducing noise, bird kill, and other environmental impacts. For photovoltaic cells, improvements may result from switching from gallium arsenide to semiconductors that have higher theoretical limits on efficiency and hence can come closer to converting all the sun's radiance that falls on the panels to electricity.

Digital electronic alternators convert DC to AC power from wind turbines and solar panels to AC power synchronized to 60 Hz suitable for grid connection, but they have two drawbacks. A drawback stems from the AC power produced by digital alternators being of poorer quality AC than what comes from turbine generators. Rapid fluctuations in wind speed and solar radiance—sometimes called their spikiness—cause ripples in the resulting direct currents. Unfortunately, alternators do not entirely remove the deleterious effects of such VRE induced fluctuations. In addition, digital alternators introduce high frequency noise into the alternating currents because they approximate the smooth sinusoidal function of AC current by stairstep approximations.

The combined effects of the direct current fluctuations' feedthrough and the alternator induced digital distortions to the sinusoidal alternating current are a challenge. Far more serious, alternators based on power electronics contribute nothing

---

[21] https://css.umich.edu/publications/factsheets/energy/wind-energy-factsheet

[22] Liu, J. et al. (2024). "Perovskite/silicon tandem solar cells with bilayer interface passivation," Nature **635**, 596.

[23] https://www.nrel.gov/news/press/2022/nrel-creates-highest-efficiency-1-sun-solar-cell.html

to grid stability. Consequently, as grid penetration of wind and solar power grows, grids will become less stable, for there will be fewer of the turbine generators discussed in Chap. 2 to provide synchronous inertia relative to the size of the grid.[24]

If inadequately controlled, they may induce unacceptable variations in frequency and voltages. Frequency deviations of more than 0.5% for even several seconds can result in generators and other equipment disconnecting—called shedding load—to prevent damage, leading to a cascading blackout. Poor voltage support can lead to voltage flutter, overheating, or worse voltage collapse.

Providing stability and reliability in the absence of sufficient numbers of turbine generators will require large compensating condensers to generate synchronous inertia and reactive power. These are electric generators, typically augmented with large fly wheels, run in reverse idling mode as motors.[25] In some locations, generators from shuttered coal plants may be repurposed to provide these functions. Such equipment is expensive, however raising the question whether it should be included as an inherent cost of installing wind and solar farms. Finding less expensive methods for countering the absence of turbine generators is an area of ongoing research focused on so-called synthetic inertial energy.[26]

The blackout that plunged Spain and Portugal into darkness on April 28, 2025, may be a harbinger of future grid stability problems. The grid was susceptible with more than 70% of its power generated from wind and solar energy at the time of the blackout. A half hour earlier oscillation in grid voltage and frequency were indicating instability. Then a substation in Granada failed, causing more than a 2000 MW power loss, grid frequency to drop, and a cascading blackout to spread over Spain and Portugal within seconds. Darkness came to commerce, industry, communications, rail transport and more. Fortunately, automated interconnects with grids in France and Morocco immediately severed to prevent the blackout from spreading internationally. Moreover, hydroelectric generators with "black start" capabilities provided power that accelerated restoration; within 20 h, power was restored to most of the Iberian Peninsula.[27]

---

[24] Ahmad, F., et al. (2023) "Dynamic grid stability in low carbon power systems with minimum inertia," Renewable Energy **210**, 486.

[25] https://www.entsoe.eu/Technopedia/techsheets/synchronous-condenser

[26] Liang, X. (2017). "Emerging Power Quality Challenges Due to Integration of Renewable Energy Sources", *IEEE Trans. Ind. Appl.* 53, 855.

[27] https://en.wikipedia.org/wiki/2025_Iberian_Peninsula_blackout

# Chapter 5
# Wind, Sun, and the Weather

**Keywords** Wind energy · Solar energy · Capacity and load factors · Intermittency · Residual demand · Backup power · Power ramping · Energy sources · Supply and demand · Energy storage · Greenhouse gas emissions · Severe weather

More than 70,000 wind turbines and 5000 solar farms span the USA, with both numbers growing rapidly.[1] The performance of wind turbines and solar panels has come a long way since their early applications. Rapidly falling costs and gains in efficiency over recent decades have accelerated wind and solar farm construction faster than would have been thought possible just a few years ago. Because wind and solar energy are weather dependent, however, the US Department of Energy classifies them as variable renewable energy (VRE). Other sources of electricity, which are independent of day-to-day changes in the weather, are classified as dispatchable or firm energy.

In proclaiming VRE's success, their growing capacity is most often emphasized, stressing that it is approaching or in some cases surpassing the capacities of nuclear or fossil fueled plants. Invariably, the capacity referred to is the name plate or installed capacity, which is the maximum power—usually measured in megawatts (MW)—that a plant can produce. For a wind farm, it is the power produced when the wind is blowing within an optimal range of speeds. For most designs, that is between 30 and 60 mph. For a solar farm, it is the power produced near noon on a cloudless summer day. Installed capacity, however, does not account for the unpredictable weather-dependent behavior of wind and solar energy. In contrast, the definition of installed capacity of a nuclear, fossil fuel, geothermal, or hydroelectric plant is independent of optimal weather conditions. It is the maximum power that their turbine generators can produce.

In comparing sources of electricity, a second parameter is as important as installed capacity. It is the capacity factor, which is a ratio of powers, usually each measured in megawatts. The numerator is the energy produced in a year, measured

---

[1] https://www.eia.gov/todayinenergy/detail.php?id=38272


E. E. Lewis, *Renewables or Nuclear*,
https://doi.org/10.1007/978-3-032-08074-5_5

in megawatt hours (MWh), divided by the 8766 hours in a year. The dominator is the installed capacity, in MW. The numerator accounts for all that happens over the course of the year of operation: how much power is needed, how many days there are of down time for maintenance, repair or refueling, the weather and other factors as well. The denominator depends on a power plant's design, and the conditions under which its maximum power is measured.

The US Energy Information Administration tabulates the annual capacity factors for the major sources of electricity each year. Figure 5.1 contains the results for 2021. They are averages over all the plants in the USA for each energy source. The largest gap is between nuclear energy and wind and solar energy, with fossil fuels and hydroelectric power between.

There are technical, economic, and meteorological reasons that lie behind the capacity factors. Solar energy has the smallest capacity factor because power production takes place only during daylight hours, the sun is dimmer in winter, and cloud cover robs solar cells of up to 90% of their power. Wind power has a somewhat higher value because it blows more evenly over the seasons. When blowing, however, much of the time it is at less than the 30 mph required for wind turbines to produce full power, and calms occasionally last for hours or even days. Due to their near-zero operating cost, wholesale electricity market rules generally result in wind and solar farms operating at the highest power that weather permits.[2]

At the other extreme of capacity factors are nuclear plants. Because they are reliable, and inexpensive to operate, they are base loaded, operating at full power 24/7 and only need to be shut down for a month or so for refueling once every

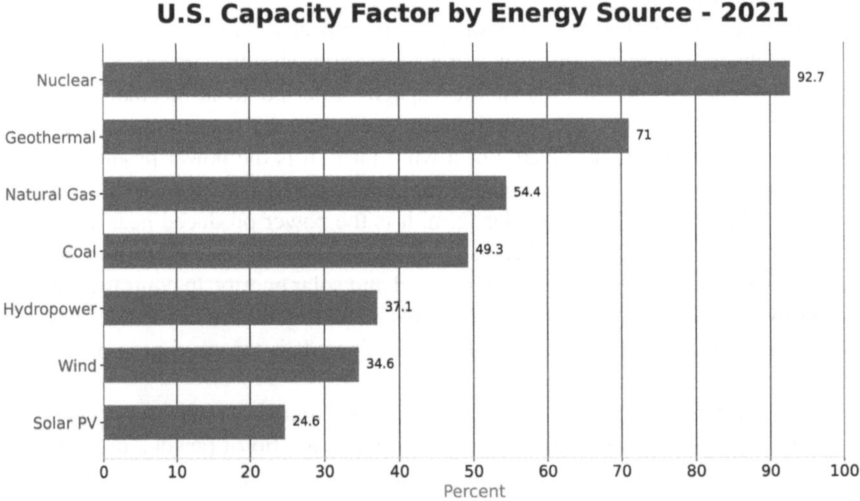


**Fig. 5.1** US capacity factors by energy source—2021. (Source: U.S. Energy Information Agency)

---

[2] https://www.rff.org/publications/explainers/ferc-101-electricity-regulation-and-the-federal-energy-regulatory-commission/

18–24 months. Geothermal plants have high capacity factors for similar reasons as nuclear plants. They, however, have no downtime required for refueling, only for maintenance and repair.

Many fossil fuel plants, particularly coal, are old, and accordingly, they operate only when needed to augment newer more economical generators. What's more, they often operate as load following and in some cases as peaker plants. Since prices of natural gas have come down in the USA, gas turbine plants have found increased use. Moreover, they produce less air pollution and only half the $CO_2$ of coal plants. Plus, open-cycle gas turbine plants—somewhat similar to turboprop jet engines—can be designed to change power level more rapidly than the other thermal power plants, making them very useful as peaking plants and in other situations were ramping up or down in power is needed.

Hydroelectric plants' relatively low capacity factors don't reflect their value very well. The water turbines located at the base of dams, typically are sized larger than they could run year-round without lowering their reservoirs to unacceptable levels. The turbines are sized to supply substantially higher levels of power for days or weeks to augment other sources of electricity when they are inadequate to meet demand. Water driven turbines can ramp from zero to full power much faster than those of fossil fuel or nuclear plants. Consequently, they are perfect for emergency situations where another power plant has suddenly shut down, and a large influx of electricity must bolster the grid quickly to avert a blackout.

Installed capacities and capacity factors must be considered together in comparing different forms of electricity generation. Otherwise, the results may be quite misleading. Comparing nuclear, wind, and solar plants, for example, all with 1000 MW of installed capacity, leads to those plants producing 927, 346, and 246 MW of power, respectively, when averaged over a year. Yet, even comparing capacity factors based on yearly averages misses the primary difficulties that arise in integrating wind and solar energy into an electric grid. The intermittency of VRE is what most challenge a grid's stability and balance. As detailed in Chap. 2, stability and balance must be maintained if grid failures resulting in blackouts are not to result.

***

Maintaining a grid in balance, with supply always equal to demand is essential, for electricity must be used immediately upon generation. Turbine generators supply dispatchable electricity on demand. In contrast, wind and solar power is uncontrollable, dependent on the weather, the time of day, and the season. Comparison of a week's worth of power versus time illustrates the disparities between VRE supply and societal demand for electricity. These must be resolved for an electric grid to function. For a week, Fig. 5.2 shows typical wind and solar supply profiles, while Fig. 5.3 shows a typical electricity demand profile that they must be made to supply. For the foreseeable future, the amounts of electricity that can be stored in batteries or by other means is minuscule compared to that needed to resolve these differences between these supply and demand profiles.

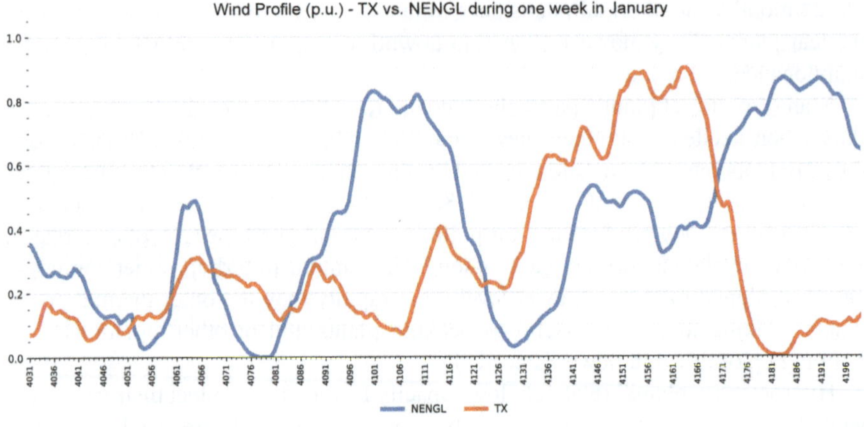


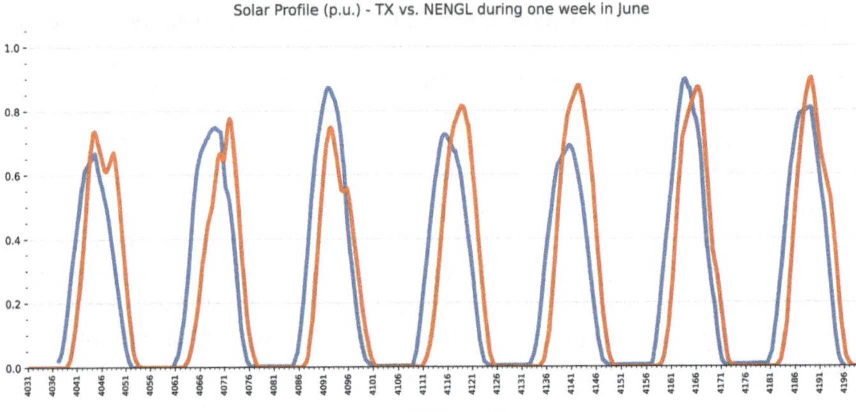


**Fig. 5.2** Wind and solar energy profiles. (Source: MIT Energy Initiative)

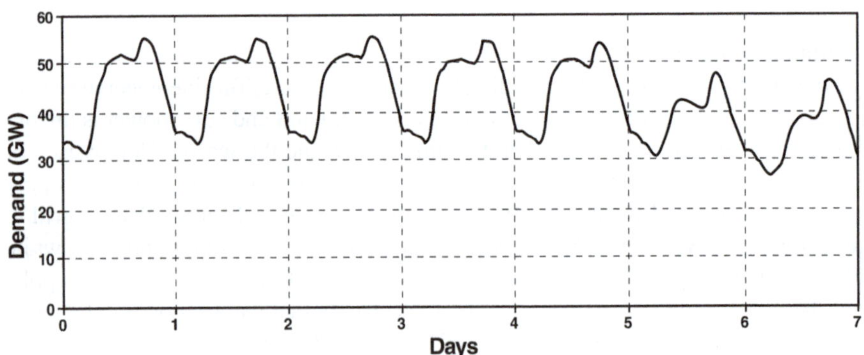


**Fig. 5.3** Typical demand profile for 1 week. (Source: Adapted from DoE)

Figure 5.2 illustrates wind and solar power for a winter week and a summer week, respectively,[3] because that is when their power tends to be most abundant. Wind intensity is erratic and difficult to predict. However, it tends to blow somewhat evenly across the seasons. Regardless of season, however, wind supply may drop to nearly zero, causing wind turbines to stand idle for the duration of the calm for hours if not days. By contrast, in the absence of clouds, solar output is as predictable as the sun's intensity, rising at dawn to a maximum at about noon and the falling to zero at dusk. Overcast skies would cause the daily solar supply peaks in Fig. 5.2 to drop to as little as 10% of the values shown. Even on deserts such as Death Valley in California, nearly 20% of days are overcast,[4] and cloudy conditions may last for several consecutive days.

Figures 5.4 and 5.5 illustrate the challenges that grid operators face in dealing with the hourly and diurnal intermittency of wind and solar energy. Each displays a week's worth of total and residual demand. The upper blue curves are the total or aggregate demand, while the lower red curves are the residual or net demand. We obtain residual demand by subtracting the VRE supply from the aggregate demand. The residual demand must be supplied by dispatchable power plants—fossil fueled, nuclear, hydroelectric or instantaneously available stored energy to make total supply equal to the aggregate demand. The figures detail two situations from a yearlong study during which VRE supplies roughly 40% of annual demand.[5]

Figure 5.4 is for a spring week during which low demand allows VRE to supply about 70% of the electricity, when averaged over the week, with the VRE produced primarily from wind turbines. The residual demand in Fig. 5.4 is nearly equal to total demand at times when the wind is nearly still, meaning that dispatchable power must provide nearly all the electricity. In strong winds, the residual demand drops to near zero, which occurs twice early in the week, and then a few days later it hits zero for several hours. During those hours, the strong winds generate more power than there is demand, and thus, wind turbines must be curtailed, and the dispatchable power plants shut down.

Figure 5.5 shows the total and residual demand for a high demand summer week during which cloud cover is insignificant, and the VRE supply is predominately from solar energy, supplying roughly 50% of the electricity, when averaged over the week.[6] The pattern of residual demand is strikingly different from Fig. 5.4. Each day at dawn solar generation ramps up from zero, providing nearly all needed energy, with wind providing the remainder. Thus, during the day, residual demand drops to zero because solar farms are generating far more energy than total demand, and the VRE must be curtailed. At dusk, solar generation drops precipitately to zero. At that time, sources of dispatchable energy must ramp up quickly to replace all the solar supply through the night and then shut down just as rapidly with the dawn.

---

[3] Tapia-Ahumada, Karen D, John Reilly et al., (2019) Deep Decarbonization of the U.S. Electricity Sector: Is There a Role for Nuclear Power? MIT energy Initiative & Joint program on the Science and policy of Global Change.

[4] https://www.bestplaces.net/climate/zip-code/california/death_valley

[5] Western Wind and Solar Integration Study, NREL & GE Energy, May 2010.

[6] California Energy Storage Alliance and LS power analysis of CAISO OASIS data 2016.

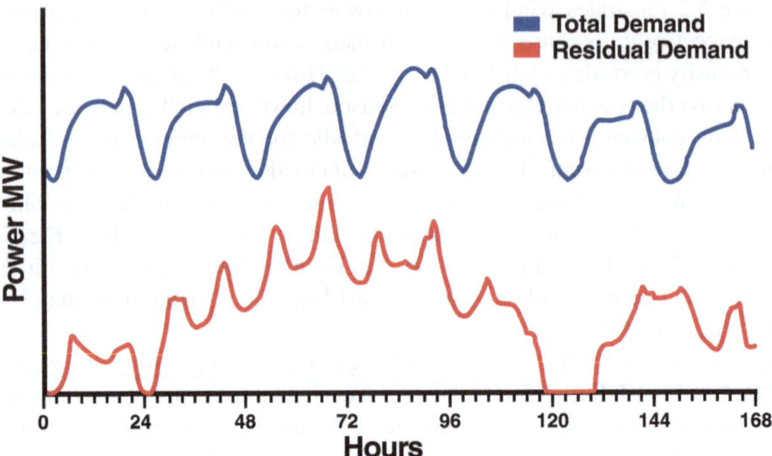

**Fig. 5.4** Impact of about 38% Primarily Wind VRE in the Western United States. (Source: Adapted from NREL)

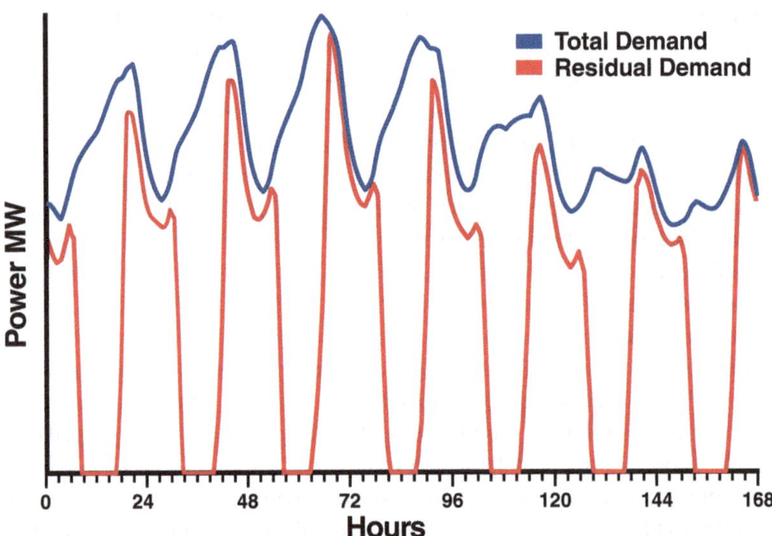

**Fig. 5.5** Impact of about 50% primarily solar VRE in the Western United States. (Source: Adapted from NREL)

Figure 5.5 is an extreme extension of the so-called duck curve.[7] It describes situations in which solar energy is dominant but insufficient to reduce midday residual demand to zero. The residual demand curve dips steeply in the middle of the day and rises rapidly in the evening as the sun sets, thus creating the duck's belly and neck.

---

[7] https://www.energy.gov/eere/articles/solar-energy-and-duck-curve

No matter whether wind or solar energy is dominant, dependence on large shares of VRE confronts the balancing authorities who operate electric grids with dire choices. A great deal of dispatchable energy is needed to counter VRE's uncontrollable intermittency. Hydroelectric plants can ramp from zero to full power very quickly and shut down just as rapidly, making them capable of coming closest to following the jagged edges and sudden changes in magnitude of the red residual demand curves. Hydroelectric power, however, is in short supply over much of the USA.

Proponents of battery storage point to it as an approach to storing electricity from when VRE is producing more than needed for when wind and solar power production is inadequate. Indeed, very large banks of lithium-ion batteries are being built, approaching 1000 MW of power, the power of a large nuclear reactor.[8] However, at full power, they discharge completely in only 4 hours. They are useful for smoothing the minute-to-minute fluctuations of VRE power and decreasing the power ramp rates required of dispatchable energy sources as the sun rises and sets over grids heavily dependent on solar energy. If the battery bank were used to compensate to the loss of solar power through a 12 hour night, it could operate only at a third of its rated capacity. If the batteries are discharged and recharged daily in such a manner, however, the entire battery bank would need to be replaced about every 2 years,[9] even while the expected lives of wind turbines and solar panels are about 30 years. Moreover, at present there is no known method for storing enough energy to compensate for cloudy skies or calm winds lasting longer than a few hours.

Efficient nuclear and combined cycle gas fired plants designed to provide steady state baseload power may provide sufficient flexibility to compensate for much of the relatively slow rises and falls in blue lines of total demand. However, they generate electricity from thermal energy. Consequently, thermal stresses, induced by rapidly changing temperatures, severely limit the rates at which they can ramp up and down in power to track the steeper peaks and troughs of the red residual demand curves. Such thermal plants also become inefficient and expensive to operate at much less than full power, and most have substantial minimum powers below which they cannot operate.

Such circumstances severely limit balancing authorities' choices. They may need to severely constrict VRE's share of generation, storing as much of the excess as possible to allow nuclear and fossil fuel plants to operate continuously at substantial fractions of full power, without starts and stops or excessive ramp rates. Even then, VRE generation exceeding 20 or 30% grid penetration necessitates fast-startup open cycle gas turbine or piston engine generators to deal with difficult to predict rapid changes in wind and solar generation.

The backup energy for VRE that fast start-up plants provide is far more expensive than base load generation. Moreover, fossil-fueled ramping operations produce nearly twice as much carbon dioxide per MWh of electricity produced as when the

---

[8] http://www.quantistry.com/blog/the-largest-batteries-in-the-world

[9] http://www.tek.com/en/documents/technical-brief/lithium-ion-battery-maintenance-guidelines

same plants operate steady-state at near full power.[10] The amounts of carbon monoxide, nitrous oxides and other air pollutants are also far in excess of steady-state
operations. Finally, for any dispatchable power plant—whatever the kind—frequent
ramping operations increase maintenance costs and the number of unscheduled
repair shutdowns, as well as substantially shorten plant life.

On grids where there are plentiful hydroelectric resources, the challenge of providing residual demand is greatly reduced. In Spain, for example, the abundant production of solar energy most often results weekly profile that strongly resembles
Fig. 5.5. At night, electricity is produced by the combination of hydroelectric and
nuclear power. The hydroelectric plants are shut down from dawn till dusk, while
the reactors are reduced to as low a power as is practical. Nevertheless, daytime
generator of solar power exceeds demand, but usually it can be used to replenish
hydroelectric reservoirs and exported to the grids of neighboring counties. Thus, the
grid is kept in balance, but stability may become a challenge, an issue to which we
shall return.

The dimming of the winter sun presents a seasonal challenge to heavy dependence on solar energy in the northern USA, Europe, and other regions far from the
equator. For nowhere near, enough energy can be stored in summer and saved for
months to compensate for the sun's dim winter rays. In Los Angeles, the ratio of
winter to summer solstice solar intensity is 0.38. Farther north, in Minneapolis, it is
0.21. Thus, even if solar farms can supply enough energy to meet summer power
demand, in winter they will supply much less, often made worse by overcast skies.

Figure 5.6 illustrates the seasonal solar conundrum. It shows the interacting
effects of cloud cover and season on solar energy in a temperate climate.[11] Both
curves are for partly cloudy days as indicated by the jagged crevasses in the power
profiles. These may be smoothed if the power output from several solar farms
located some distance apart are averaged or if adequate battery storage is available
to eliminate the curves' cervices. In temperate climates, dispatchable sources of
electricity must supply nearly all the power on cloudy days and through the winter months.

What are the greenhouse gas emission consequences of employing VRE power
on electric grids with various combinations of fossil fuel and nuclear energy? The
mass of $CO_2$ emitted per MWh of electric energy, employing fossil fuel, VRE, and
nuclear generators provides a useful comparison. Figure 5.7[12] indicates the results
in blue. These are "life cycle" numbers, including the greenhouse gas emissions that
result from the production of steel, concrete and rarer materials used in power plant
construction as well as those incurred in mining, fabrication, and plant
construction.

---

[10] http://euanmearns.com/co2-emissions-variations-in-ccgts-used-to-balance-wind-in-ireland/

[11] Baetens, R., et al., (2010). "The Impact of Load Profile on the Grid-Interaction of Building
Integrated Photovoltaic (BIPV) Systems in Low-Energy Dwellings," *Journal of Green Building*
5(4) 137.

[12] https://docs.nrel.gov/docs/fy21osti/80580.pdf

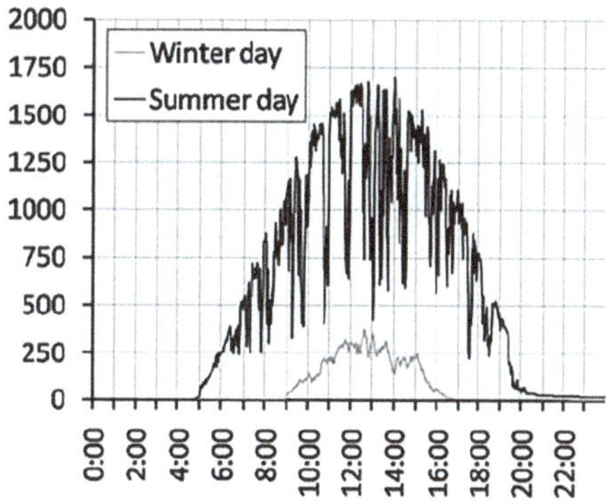

**Fig. 5.6** Solar PV power vs. time for a temperate climate. (Source: *Journal of Green Building*)

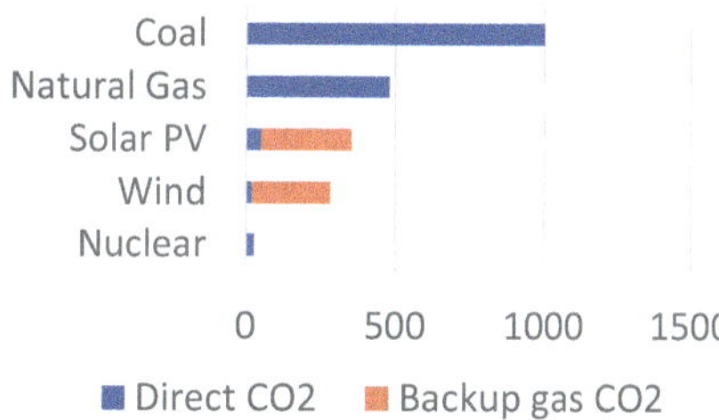


**Fig. 5.7** Greenhouse gas emissions in $CO_2$/MWh from power generation. (Source: Adapted from NREL)



Replacing coal with gas cuts $CO_2$ emissions in half, but further gains with VRE are marginal because of the need for fast-ramping dispatchable backup. The $CO_2$ emissions from backup gas-fueled plants are indicated in orange. They are larger than might be expected. They result because the small capacity factors of VRE mean that more than half of power production must be from backup plants. Moreover, backup plants' poor combustion during frequent ramping produces sustainably more $CO_2$ than if they operated at steady state power. If nuclear energy were to carry the full load, emissions would be miniscule, only including those for the manufacture and construction of equipment. Not included in the figure are the greenhouse gas emissions from methane leaking to the atmosphere. Estimates vary, but methane is between 30 and 80 times as potent as $CO_2$ as a greenhouse gas.[13] Thus, just the 1% leakage would almost wipe out the advantage of replacing coal with natural gas to fuel power plants, and greatly diminishes the value of VRE as well.

***

In the coming decades, technological advances detailed in the chapters that follow may mitigate the difficulties of VRE penetration of power grids illustrated by the foregoing figures. Carbon capture and sequester may allow fossil fueled plant to be decarbonized, while also retaining ramping capabilities. Advances in energy storage technology should be helpful in smoothing the hour-by-hour peaks and

---

[13] https://www.unep.org/news-and-stories/story/methane-emissions-are-driving-climate-change-heres-how-reduce-them

troughs in residual demand, such as shown in Fig. 5.4, and larger amounts of solar energy may be stored from midday until evening to level somewhat the residual demand shown in Fig. 5.5. Advanced nuclear reactors may have improved capabilities for ramping up and down in power and for operating at low power: Naval nuclear propulsion technology may be adapted for power production. Thermal energy stored from high-temperature reactors running at full power may be used to ramp steam turbine generators up and down in power. Pairing nuclear and VRE generation to simultaneously produce electricity and hydrogen, process heat, or energy storage also has attractive possibilities.

Beyond typical hour by hour and day by day variations in wind and solar power and the predictable seasonal changes in solar energy, balancing authorities must deal with severe albeit rare weather events. In the USA, the polar vortex of 2019 serves as an example. A gale brought artic air and a snowstorm to a region stretching from Minnesota into Iowa and beyond that was dependent on wind farms for half of its electricity. Then, for the last 4 days of January, the worst polar vortex in more than 20 years ensued. As the wind died away, wind power slid to roughly 2% of what it had been,[14] while temperatures plummeted to as low as −33 °C (−28 °F) in Minneapolis and to −50 °C (−60 °F) in some rural locations. Schools closed, flights were canceled, postal service stopped, and water mains broke.

The energy sector was in crisis mode. Wind turbines shut down automatically below −29 °C (−20 °F) to prevent damage, and in fact consumed electricity to prevent lubricants from freezing or metal machinery from fracturing. Frozen piping and other maladies forced some fossil fueled plants offline. Natural gas intended to back up wind power was scarce, since law prioritized the need for home heating over electricity production. Gas usage skyrocketed, causing a statewide request to turn back thermostats to 17 °C (63 °F). A blackout would be disastrous under sub-zero conditions. Electric furnace controls would have failed, causing widespread loss of home heating across the region, and communications would have been disrupted.

Thankfully, a blackout was narrowly avoided by securing adequate power to meet demand: Minnesota's two nuclear power plants didn't falter, providing 14% of the needed power. Retired coal plants were brought back online and accounted for 45% of the needed electricity, and natural gas plants for an additional 26%.[15] Energy imports from Canada and from lesser affected neighboring states provided the remainder.

The challenges encountered when solar power supplies a substantial fraction of the energy tend to occur under different conditions. For example, an unprecedented triple digit heat wave combined with severe wildfires brought the urgency of fighting climate change to California in August of 2020.[16] There, effects on the electric

---

[14] https://energynews.us/2019/02/27/wind-turbine-shutdowns-during-polar-vortex-stoke-midwest-debate/

[15] Ibid.

[16] https://www.utilitydive.com/news/california-releases-final-root-cause-analysis-of-august-rolling-blackouts/593436/

grid serving 80% of the state were frightening. The statewide effort to accelerate the transition from fossil fuels to renewables had resulted in the decommissioning coal and gas-fired plants, and public sentiment had forced the closing of one the state's two nuclear power plants. More than a third of in-state electricity already came from renewables, and much of that from solar energy.

Air conditioning use was driving demand for electricity up by 50% or more from that of a typical summer day. While California's geothermal and its remaining nuclear power plant ran at full bore, the heat caused some gas-fired plants to go offline, and wildfire smoke was diminishing the output from some solar farms. Drought and low snowpack in the mountains the previous winter diminished hydro-electric power within the state. Thankfully, the contracted importation of the third of the state's power, much of it coming from hydroelectric generation in the Pacific Northwest, continued unabated. But making matters worse, efforts to increase purchases of electricity generated in surrounding states fell short. The heat wave spread over the West, and they had little to spare.

As could be expected, the crunch came in late afternoon and lasted into early evening. The setting sun's intensity plummeted, while electricity demand only followed suit some hours later. In normal circumstances, fast starting gas-powered peaking units would pick up the slack, but they already were running at full power to meet the record-breaking demand. Nor did it help that California's considerable wind power capacity dwindled at that worst of times. Neither were grid operators' calls for Californians to voluntarily reduce their power use during those critical hours sufficient to stem the crisis. To prevent a cascading blackout over much of the state that would likely last for many hours, if not days, the California Independent System Operator imposed rolling blackouts, leaving hundreds of thousands of homes and business in the dark. The blackouts ended when wind generation finally picked up.

***

The foregoing discussion focuses on technology, touching on what the future may hold in the way of technological advancement to deal will challenges stemming from the intermittency of wind and solar energy. Economics, however, plays and equally important role, for the technology must be affordable. On power grids with large penetrations of VRE, the dispatchable power generators, whether they be fossil, nuclear, hydroelectric, geothermal, or energy storage must also operate in variable modes to complement VRE generation, such that the sum of the VRE and dispatchable power supplies is exactly equal to the varying daily cycles of total demand shown in earlier figures. With rare exception, the capital costs of building decarbonized power plants—no matter whether they generate VRE or dispatchable energy—greatly outweigh the fuel and other operational costs. Consequently, the price that must be charged for electricity—termed the Levelized Cost of Energy (LCOE)—is roughly the capital cost divided by the total megawatt hours of electricity generated during the life of the plant.

On a grid with little or no VRE, the more efficient plants are base loaded, operating near or at full power 24/7 with capacity factors close to one. Those plants, however, must be used instead to compensate for the variable output of wind or solar energy. Their complementary operation entails operating a low power levels and frequent ramping up and down in power, which thus averages much less than full power. In addition, such operations require more down time for maintenance and repair. These two factors result in substantial reductions of the plants' capacity factors. For example, VRE grid penetration resulting in cutting the capacity factors of dispatchable energy plants in half would necessarily nearly double their LCOEs. Thus, as the share of VRE on a grid increases, the cost of electricity to its consumers will increase, even though the costs of building wind and solar power plants may be plummeting.

The economics of energy storage is somewhat different. Again, capital cost most often dominates, and revenue to recover capital costs is proportional to the number of buy–sell cycles over the life of the facility. While 24-h arbitrage may be profitable, storing energy for days, weeks, or longer durations to compensate for diurnal or seasonal variations or for use in infrequent emergency situations would require exorbitant storage pricing. Chapter 16 examines these and other factors related to the economics of electric grids in more detail.

# Chapter 6
# Traversing the Landscape

**Keywords** Wind energy · Solar energy · Land requirements · Public opposition · Wind and solar farms · Esthetic impacts · Shuttered coal plants · Geothermal heat pumps · Opinion polls · Offshore wind energy · Grid deregulation

It was the first trip I'd made in more than a decade from suburban Chicago to Central Illinois for a visit with my late wife's sister and her husband. I'd always enjoyed spending time with them, but our more recent visits had all been at my place. For this particular trip, however, I had two motives beyond just having a pleasant reunion. I wanted close-up views of the wind and solar farms that they had told me were going up near their rural property, and I was curious to see how the geothermal heat pump system worked that they had installed to both heat and cool their home.

Following Interstate I55, which had replaced the famed Route 66, I passed a familiar site some 60 miles southwest of the city. On my left, two containment buildings of the Braidwood Nuclear Power Plant were just visible, peaking over the treetops. Some miles further was the exit on the right where decades ago I would leave I55 and follow the road to the Dresden Nuclear Power station where my consulting for Commonwealth Edison had taken me. So different the landscape had become in recent years. Off and on for the next hundred miles and more, wind turbines spread across the expansive fields of corn and soybeans that are the mainstays of Illinois agriculture. Though only visible from closer vantage points, there were also fields filled with row upon row of solar panels. It seemed I was proceeding on an unguided tour of the future of variable renewable energy (VRE) as wind and solar energy are designated by the Department of Energy.

The trip was altering my views of where the growth of VRE was taking place. Much had been written of the wind energy potential of the sparsely populated Great Plains, and the need for greatly expanded construction of high-voltage transmission lines to carry that power to population centers east of the Mississippi. But here it struck me that something different was happening. In Illinois, more than twenty coal fired plants, several which fed electricity to the Chicago metropolitan area, have been shuttered in little more than a decade. Some closures were before the concern with greenhouse gas emission had skyrocketed. The plants emitted amounts



E. E. Lewis, *Renewables or Nuclear*,
https://doi.org/10.1007/978-3-032-08074-5_6

of sulfur dioxide, fine particle soot, and nitrous oxides that failed to conform to Environmental Protection Agency pollution abatement requirements and were forced to close. As a result, the counties surrounding such sites were proving to be opportune places to locate VRE farms. Expensive new high-voltage transmission lines need not be built. Substations built close to wind or solar farms could transmit power at lower voltages over relatively short distances to existing high-voltage transmission lines that connected abandoned coal fired plants to Chicago and St. Louis.

As I drove on past fields of corn, dotted with wind turbines, I recalled that nearly half of the corn grown in the USA becomes feedstock for producing ethanol, a bio-fuel. More interesting, to produce the same amount of energy, growing that corn requires nearly twenty times the land area as a wind farm, and about 100 times that of solar panels.[1] Thus, the land sprawl caused by growing corn for ethanol is far greater from that of wind and solar farms. But the corn doesn't alter the landscape significantly while VRE certainly does. Yet now with the value of ethanol in curbing climate change being disputed, it seems that increasing the food supply in less fortunate parts of the world would be a better use of the plentiful corn crop. Unfortunately, humanitarian concern often doesn't win over politics, and the well-entrenched ethanol lobby would be a formidable foe of limiting ethanol production. A first step for anyone wanting to be US President is doing well in the Iowa caucuses, and ethanol subsidies are what keep the price high of the corn grown on Iowa farms.

The size of the turbines I passed varied, with the largest occupying the newer wind farms. There was incentive for building ever larger turbines, for higher off the ground, their blades incurred less turbulence and steadier winds. Greater height also capitalized on the economies of scale, for longer blades produced more megawatts (MW) per turbine. The growth of these behemoths, however, requires placing them further apart so that the wake of the wind from one does not curtail the efficiency of its neighbors. Consequently, the MW of electricity generated per square mile of wind farm remains about constant.

As I drove on, I found myself thinking about old fashioned windmills, the kind my parents used to have me count for entertainment on long road trips on two lane highways before the interstate system came in to being during the Eisenhower administration. They were small and seem even smaller when pictured next to a modern wind turbine as shown in Fig. 6.1. An American invention of the mid-nineteenth century, the windmills were perfectly matched to needs of the farmers of the Middle West and Great Plaines: an intermittent source of water, coupled with storage. Water pumped by a windmill was stored in a large open tank or trough at its base. Cattle could drink whether the wind was blowing or not, and only rarely, during wind droughts, would a hand pump need to be used. Railroads also placed

---

[1] https://www.cleanwisconsin.org/wp-content/uploads/2023/01/Corn-Ethanol-Vs.-Solar-Analysis-V3-9-compressed.pdf

Fig. 6.1 Windmill and wind turbine. (Source: T. Tramm)

the same variety of windmill along their lines to gather water needed to replenish steam engine boilers.

Such windmills were ingenious mechanical devices, with the vane connected to the sails, which we would now call blades, in such a way to keep the windmill facing the winds when they were light or moderate, but turn its face parallel to strong winds, thus preventing damage to the sails. Their use diminished in the 1920s as the availability of grid electricity spread from cites into rural areas. However, in more remote locations beyond the reach of electric utilities, the old-style windmills were often redesigned to power small electric generators, providing at least intermittent electricity to many farms. These small wind turbines were the forerunners to the megawatt machines that I'd been viewing.

I turned off I55 and drove along country roads past Easton, Illinois, the village nearest to where my in-laws live. I had close-up views of both towering wind turbines and row upon row of solar panels. I stopped so I could hear the swishing sounds of the turbine blades and got out to stretch my legs and take a closer look up at the turbine's blades and nacelle. It takes being up close to appreciate the size of such behemoths, with blades the length of a football field, and tower heights nearly double the blade lengths. Equally impressive was what I could not see. Buried beneath 2 or 3 feet of soil would be turbine's foundation, a massive cylindrical slab of reinforced concrete close to 10 feet thick with an 80 foot diameter, and weighing more than the turbine itself.

The motion of the blades was deceptive. They rotate at about twenty revolutions per minute, which from a distance makes them look to be moving quite slowly. But that's only true close to the hub; typically, the blade tips spin along at 100 miles per hour or more. As I watched the blades' rotation, I thought of a graph depicting another aerodynamic phenomenon: It showed the interference of air flow caused by the turbine's tower resulting in a power dip of nearly one-third each time the tower interferes with the wake of a turbine blade, about once per second.

Sitting atop the turbine's tower is the garage-sized nacelle, packed with mechanical, hydraulic, and electrical equipment all tied together and controlled by on-board computers. The simplified diagram of Fig. 4.2 could not do justice to the complexity of the maze of equipment. Packed around the gear box and generator, dozens of sensors monitoring speeds, voltages, temperatures, stress and vibration levels, and more. The data they generate is processed by on-board computers, and messages are sent to wind farm headquarters when maintenance or repair is needed. A cooling system must also be squeezed into the nacelle to prevent over heating of the gear box and generator. Electric power must be brought from the grid to the nacelle to keep the cooling system, computers, and other equipment running. Even when the turbine shuts down due to excessive wind speed or subzero temperatures, the grid must continue to supply power to keep essential equipment running, and in the case of subzero temperatures to provide heat to keep gear box lubricants from freezing.

I returned to my car and soon turned off the county road and completed my trip at the end of 200-yard drive leading to my in-laws' house. After a warm welcome and a delicious dinner, our discussions turned to the wind farm that was to be located on the other side of the county road that I had just exited. My brother-in-law unfolded a map that was the initial proposal for the farm to show me how it related to their property. The farm would consist of 52 wind turbines. I found its checkered layout fascinating. Apparently, many landowners didn't want wind turbines on their land, making the farm's outer boundaries enclose nearly twice the area needed for minimum spacing of the turbines. The wind farm's developers, of course, hold land rights to run access roads to each of the turbines. They also install a highly optimized subterranean array of power lines and junction boxes. The array collects the electricity from the turbines, where it has been synchronized to 60 Hz, consolidates it for more efficient transmission, and delivers it to the electric substation. At the substation, the electricity is increased to higher voltages and transmitted to the grid.

The wind farm, along with its substation, is situated on the corn and bean fields on the opposite side of the county road bordering my in-law's property. They had built their home in an open area next to a ravine that made the surrounding woods unsuitable for agriculture. Visual pollution, swishing sounds and bird kill that would likely accompany the wind farm caused them concern. However, five active eagle nests and a breeding area of Illinois chorus frogs, an endangered species, occupied the wooded area of my in-laws property. Thus, state law required the wind turbines to be set back a considerable distance back from the other side of the road. More important, after the map had been drawn their neighbor across the road decided against having a turbine on his land. Consequently, the nearest turbine would be located at a more distant location. From my in-law's home, only the sail tips of the farm's nearest turbine should be visible above their trees, and its swishing sound should be nearly inaudible.

***

The next day, we toured a wider area, first looking at the substation under construction to serve the wind farm. We then drove to the Illinois River to see the shuttered 488 MW Havana coal plant close to where the wind farm's power lines would connect to the high-voltage lines leading to Chicago and St. Louis. As we approached, the first signs of Havana were the abandoned plant's two tall smokestacks rising above the tree line. Across the road from the plant, several large ponds had been used for coal ash disposal. They now appeared as expansive grass covered embankments, some more than 20 feet high. The plant's coal supply had come down river by barge and was piled on a riverside field opposite the plant. That space had since been cleared and was now used to store wind turbine blades until trucks carried them to wind farm sites. We continued our drive, seeing not only several wind farms, but also the extensive panel arrays of solar farms.

That evening after dinner we drove a short distance from my in-laws' home to see wind energy's most visible imprint. The wind farms were nearly 20 miles away, so far that in daylight they were hardly visible. Yet, at night, they appeared as a string of blinking bright red lights stretching across the horizon. My brother-in-law described the scene as looking like an airport runway. He found the lights particularly irritating because the couple once enjoyed watching harvest moons rising above the horizon, but now the wind turbine lights encroached upon their view.

After returning from our outing, discussion turned to the obvious reasons why solar farms are much less obtrusive than wind farms. They are more compact, taking only about a fifth of the land of a wind farm of the same installed capacity. They rise only several feet above ground level, require no blinking lights, and on level terrain become invisible from only a few hundred yards away. But the solar and wind farms we viewed were similar in that they both fed electricity to a substation where its voltage was increased and connected to the grid's high-voltage transmission lines. They thus are both considered large or utility scale. There is another segment of solar power generation, discussed in Chap. 4. It consists of smaller instillations at commercial, industrial and residential locations that also derive power from solar cells. Taken together, they presently account for 43% of the solar market but are the most rapidly growing segment.[2]

* * *

The last day of my visit was rainy, which was a good time to visit the basement and see the geothermal heat pump system that freed my in-laws both from air conditioners and from the propane heating system they used to have. Heat pump systems both heat and cool, using substantially the same piping. They use much less power than electric heating because electricity is not used to heat the indoor areas directly but only to move heat from one place to another. They serve as a reversible air conditioner. In summer, they move heat out of the house, even though it is hotter

---

[2] https://www.gresb.com/nl-en/community-solar-is-the-fastest-growing-segment-of-the-us-solar-market

outside than in. In winter, they move heat into the house, even though it is warmer inside than out.[3]

Window units designed to control the temperature in a single room are the simplest, exchanging heat between the room and outdoors. Duct work installed to provide central heat and air conditioning can also be made part of a heat pump system for houses or larger buildings. Too bulky to fit in a window, the heat pump likely will be located outdoors. For apartment buildings with centralized heat and air conditioning, placing the heat pump on the flat roof is increasingly popular. With the foregoing arrangements, the heat pump exchanges heat between indoor and outdoor air temperatures, but unfortunately its efficiency decreases as the temperature differences increase. Hence, the pump must work hardest—and consume the most electricity—on the hottest and coldest days of the year.

If land is available and the installation cost not too onerous, then the shallow geothermal system diagramed in Fig. 6.2 is more energy efficient. It exchanges heat with the earth six or more feet below the surface where the temperature is virtually unchanged year-round, midway between summer and winter outside temperatures. An underground piping loop, such as Fig. 6.2 illustrates, heats water in winter, and the heat pump further increases its temperature to heat the house. In summer, the piping loop cools water, which the heat pump then further decreased its temperature to cool the house. Thus, the underground system had much improved year-round efficiency.

My in-laws' geothermal heat pump is more elaborate than the one diagramed. Instead of exchanging heat with the subsurface soil, water comes from their 200 foot deep well. It is a once-through system. In it, the water flows directly from the

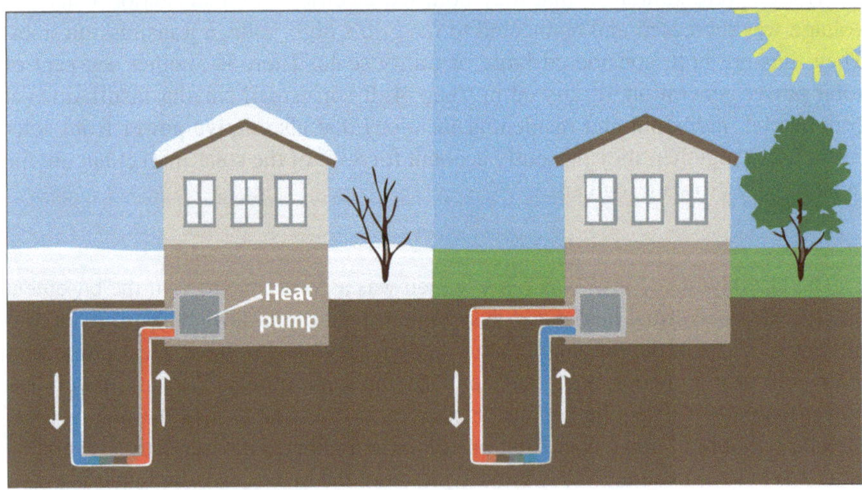


**Fig. 6.2** Geothermal heat pump system. (Source: Adapted from EPA)

---

[3] https://www.energy.gov/energysaver/heat-pump-systems

well to the heat pump. It then drains to the ravine near their house, thus eventually, returns to the aquifer beneath their land. My brother-in-law pointed out that since heat transfer between flowing well water and the heat pump is more effective than from soil to the water in a piping loop and then to the heat pump, the once through system is more efficient, even while less piping is needed. The system of pipes and valves in their basement was interesting, albeit somewhat complicated. It was configured to separate cold water for drinking before it arrives at the heat pump, and the hot water heater was integrated into the heat pump system to minimize the need for electricity. The trip to the basement was a welcome digression from our ongoing discussion of VRE.

After lunch, we turned our attention to analyzing local newspaper clippings and web sites describing the Moraine Sands Wind Farm that will boarder their road. Each turbine will have an installed capacity of 4.2 MW.[4] They are huge; their blade tips circling to a height of nearly 700 feet,[5] taller than the 555 feet Washington monument! Just as striking is the land area each turbine must occupy so that its wake does not interfere with neighboring turbines. The required land area increases as the square of the turbine height. Each of which will need about one-fifth of a square mile. Since the farm is planned to have 52 turbines and hence must occupy more than 10 square miles. However, the turbines will spread over a wider area than that, since no turbines will be located on a number of land parcels because the terrain is unsuitable, the landowners don't want them close to their homes, or the impact they might have on farming operations.

I started wondering how much land must be devoted to wind farms to equal the power of the two reactors at the Braidwood plant I had driven by. Taken together, they have an installed capacity of 2400 MW. While the efficiency of wind turbines has been increasing with their size, the land area they require has grown as well. As a result,[6] the maximum power wind turbines produce per square mile has remained roughly constant at about 10 MW/mi$^2$. That means for wind farms to equal the power of the Braidwood plant, averaged over a year, they would need to occupy 240 mi$^2$. That is just about equal to the land area of all of Chicago. An analogous calculation for Solar panels yields a much smaller area. Only about 110 mi$^2$ would be required to equal the power of the two reactors averaged over a year. Solar panels, however, cover the ground nearly completely. Thus, unlike the areas covered by wind farms, the land cannot be used for farming or grazing livestock. Still, either wind or solar farms require a lot more land than taken by the Braidwood reactors, which together occupy 2 or 3 square miles.

***

On my drive home, I thought about the rebellion growing among communities close to the vast tracts of land that VRE farms must occupy to meet decarbonization

---

[4] https://www.mortenson.com/projects/moraine-sands-wind-project

[5] https://us.vestas.com/en-us/products

[6] https://sciencing.com/much-land-needed-wind-turbines-12304634.html

goals. Some distance to the east of my in-laws' home a 50-turbine farm is being proposed for a location near the central Illinois town of Monticello. Objections include visual pollution of bucolic landscapes, blinking lights, real estate values, bird kill, noise and more. After 13 nights of hearings, the County Zoning Board's five members voted unanimously against recommending a permit for the wind farm. One argument made was that 70% of the land on which the wind turbines would be located was not locally owned.[7] That made me wonder what the percentage would be for the Moraine Sands project across the road from my in-laws.

In Iowa, 16 of the state's 99 counties have passed prohibitive ordinances or moratoria against wind farms,[8] most enacted in the last few years. Those restrictions will make between half and three quarters of potentially suitable areas unavailable. The numbers of similar county bans are growing in Ohio, Michigan and other middle western states. And it's not only in the Midwest where resistance is growing. From the largest state to the smallest—from Texas to Road Island—opposition to wind and solar farms is intensifying.[9] Nationwide there have been more than three hundred local governments placing bans or moratoriums on utility scale VRE projects.[10] A tabulation of the number of local restrictions being kept by the journalist and author Robert Bryce, shown as Fig. 6.3. It indicates the rising tide of objections to expanding the land covered by wind turbines and solar panels.

There seems to be a disconnect. National opinion polls strongly support the expansion of renewable energy as do many state governments. In Illinois, the legislature passed a bill stating that by 2030, 40% of electricity should come from renewables. If the national polls are examined more closely, however, they generally solicit only brief responses, asking general questions such as do you support wind and solar energy development to stem climate change? When more detailed questions are asked regarding the benefits and drawbacks of renewables, then support tends to wane.

Urbanites viewing renewables as the way to fight global warming tend to view the rural backlash as just another form of the "not in my back yard" phenomenon. Views tend to change, however, when wind farms threaten to come within daily view. In the middle west, this is happening as developers contemplate harnessing the strong and steady winds over the Great Lakes. The lakes' shores are lined with pristine vacation properties and communities with economies based on tourism and fishing. Any wind farm visible from shore would likely meet with opposition as strong as the Massachusetts lobby that defeated the Cape Wind Project, a 130-turbine farm that would have occupied 24 square miles of Nantucket Sound.[11]

---

[7] https://www.nytimes.com/2022/12/30/climate/wind-farm-renewable-energy-fight.html

[8] https://clearpath.org/reports/hawkeye-state-headwinds/

[9] https://www.wsj.com/articles/texas-clean-energy-renewables-opposition-a654a2d5
   https://ecori.org/2018-3-15-a-contentious-battle-green-energy-vs-green-space/

[10] https://www.nytimes.com/2022/12/30/climate/wind-farm-renewable-energy-fight.html

[11] https://www.reuters.com/article/usa-windfarm-idAFN2825518020100428/

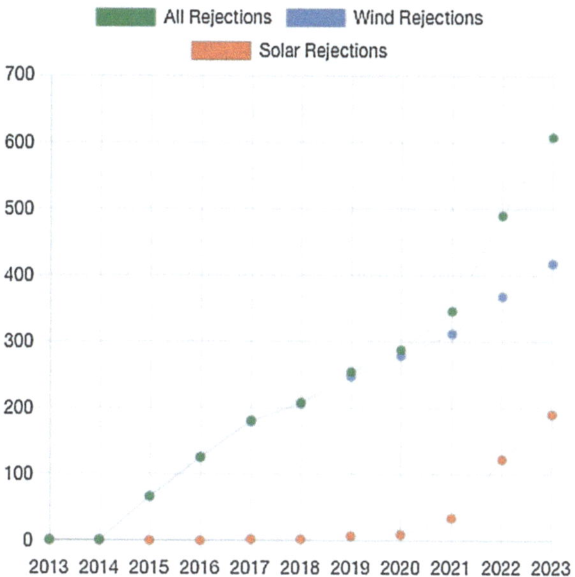


**Fig. 6.3** Number of VRE restrictions vs. time. (Source: Robert Bryce)

Imagine the uproar if a Lake Michigan wind farm were proposed with its blinking red lights visible from the skyscrapers that line Chicago's gold coast, even if only from the highest venues, such as the restaurant atop the 100-story John Hancock building. And I'd be unlikely to appreciate those lights while enjoying a walk along Northwestern's lakeshore campus. Fortunately, my in-laws recently learned that for the red lights atop the wind turbines of the Moraine Sands Wind Farm, the blinking light problem will be diminished by coupling them with motion detectors so that they will only flash when aircraft are in the vicinity. Motion detectors are likely to have little effect, however, on the frequent occasions when the landing pattern into O'Hare International Airport brings aircraft in over Lake Michigan every few minutes.

The dilute nature of VRE results in vast tracks of land being needed for wind and solar farms, and the farms often are located much further from the population centers where electricity is consumed than those I visited in Illinois. For example, how do the wind farms spread across the Great Plains transmit their energy to the population centers on the Coasts, the Gulf of Mexico or the Great Lakes? The challenge is made greater because state and local governments generally have authority over siting and construction of transmission lines. Moreover, transmission lines for VRE will rarely be confined to a single state, or to two adjoining states. At least both states have an economic interest, one the producer, the other the consumer. But if the transmission lines must pass through intermediate states, their citizenry may see no economic benefit and therefore oppose the line. It's understandable why permitting for such lines may take 10 years or more.

The process was somewhat less complicated before deregulation of electric grids in the 1990s. Earlier, vertically integrated utilities owned both power plants and transmission lines. Thus, they could simultaneously plan the locations of new power plants and the transmission lines emanating from them. For fossil fuel plants, they would also plan barge, railroad, or pipeline routes to get fuel to proposed power plants. Now in much of the country merchant generating companies own and operate the solar and wind farms, while the responsibility for building transmission lines lies elsewhere, spread between multiple federal and state agencies.

The challenges of spanning transmission lines over the vast distances between inland VRE sites and coastal population centers have made offshore wind farms an attractive alternative, for they offer several advantages over their land-based counterparts. They are typically between 20 and 100 miles offshore, far fewer miles than many overland transmission lines would need to traverse, and the myriad of hindrances caused by state and local permitting processes are avoided. Steadier ocean winds also increase capacity factors from the 30% range to as high as 50%.

Offshore wind, however, entails higher cost per MW than those onshore, even though the individual turbines are larger, some 10 MW or more. They must be anchored to the ocean floor, along with substations and transmission lines to carry power to land. The offshore projects are large. New York state recently contracted for three offshore farms totaling 4 GW of installed capacity.[12] But even with a capacity factor of 50%, the average power that they present is not much more than the 2 GW provided by the two Indian Point nuclear reactors that were recently shut down prematurely. Moreover, unlike wind turbines, the reactors didn't shut down during hurricanes, during the times that power is most needed.

Dispatchable forms of renewable energy are not subject to VRE's day-to-day variations of the weather. Moreover, unlike VRE, they do not require great expanses of land. Dispatchable renewable energy ranges from large dams with hydroelectric plants to tidal and geothermal energy. Broadly classified as hydropower, we examine them in the following chapter.

---

[12] https://www.power-technology.com/news/new-york-offshore-wind-4gw/

# Chapter 7
# The Power of Water

**Keywords** Hydroelectric power · Historical development · Dams · Hydroelectric versatility · Dispatchable power · Ramp rates · Capital cost · Long operational lives · Droughts · Tidal power · Geothermal power

Niagara Falls, shown in Fig. 7.1, are spectacular. When my wife and I first visited the falls, we understood immediately why a trip to see them had become a honeymoon tradition. The colored lights illuminating them are not easily forgotten when viewed from an evening boat tour lingering not far downstream on the Niagara River. There are, in fact, three sets of falls to marvel.[1] Horseshoe Falls, sometimes referred to as the Canadian Falls, are the largest. On the other side of Goat Island, through which runs the border between Canada and the USA, are the American Falls along with the much smaller Bridal Veil Falls. While the falls power and beauty stand out, casual tourists often don't appreciate the importance they played

**Fig. 7.1** Niagara Falls. (Source: iStock)

---

[1] https://www.niagarafallslive.com/facts_about_niagara_falls.htm

in the beginnings of hydroelectric power, or that they are unique in that without damming the river, they have been a major source of electricity for well over a century.

Following Thomas Edison's invention of the incandescent light bulb and his creating the first electric grid from his Pearl Street station in New York City, the race was on to find better ways to produce and transmit electricity. Using the power of water was foremost among them.[2] In 1880, a small water turbine generated electricity for arc lighting a storefront in Grand Rapids, Michigan. In 1882, Edison built a small dam and generator on the Fox River in Wisconsin, which supplied lighting to some nearby buildings. Edison, however, was wedded to the direct current systems he had invented, and that restriction made the transmission of electricity over long distances exceeding inefficient and thus cost prohibitive. Meanwhile, the brilliant physicist Nikola Tesla worked out the theory of alternate current (AC). He partnered with the inventor and manufacturer George Westinghouse to build the first AC hydroelectric plant, which derived its power from Niagara Falls.[3]

Their hydroelectric system didn't dam the falls. That would have been impossible for several reasons. Instead, in 1894, Tesla and Westinghouse diverted a small fraction of the Niagara River's water from upstream of the falls along a channel and then down through a tunnel and into their AC water turbine on the Niagara River Gorge near the base of the falls. Accordingly, they made use of the difference in elevations to convert the water's potential energy to electricity. Critically, using alternating current allowed them to transmit electricity efficiently from their generator to Buffalo, New York, 25 miles away, and provide lighting and power to grow the city's industrial base.[4]

More powerful hydroelectric plants on both the US and Canadian falls have supplanted that pioneering installation.[5] Today, 2.7 gigawatts (GW) and 2.2 GW of electricity are generated by the two nations, respectively. The total capacity of 4.9 GW is more than the 2.1 GW of the Hoover Dam. To accomplish this, however, 60% of the Niagara River is diverted through tunnels around the falls, leaving only 40% for tourists to enjoy. However, the reduction in flow over the falls is hardly noticeable to visitors, even more so because so few have ever seen the falls before their flow was restricted. Late at night, after the colored lights on the falls have been turned off, and after the close of the tourist season as well, even more water is diverted to increase electricity production.

There is a flip side to discontent over water diversion. Although the falls may seem a little less spectacular, the decreased flow over them slows their rapid erosion from the water's force. In 1954, a section of the US side of the falls broke free, dumping 185,000 tons of rock into the Niagara River Gorge, somewhat diminishing

---

[2] https://www.nps.gov/articles/5-the-origins-of-hydroelectric-power.htm

[3] Ibid.

[4] Ibid.

[5] https://www.niagaraparks.com/visit-niagara-parks/plan-your-visit/niagara-falls-geology-facts-figures/

the falls allure. Without the water diversions to the power plants, however, rock falls large and small would be expected to occur far more frequently. The Niagara River and human ingenuity provide a unique combination of water falls and dam-free power plants successfully wrapped into one. And even with erosion, hopefully the falls will continue to provide both recreation and power for decades if not centuries to come.

***

Figure 7.2 shows how a hydroelectric plant converts the potential energy of the water stored at a high elevation to kinetic energy of the water streaming through the turbine's penstock at a lower elevation. The impact of the water on the turbine causes its shaft to spin at 60 Hz. The generator then converts the shaft's mechanical energy to alternating current electricity. The power of the hydroelectric plant is proportional to the height of the reservoir's water level above the turbine and increases with the rate at which water is allowed to flow through it.

Unlike wind and solar energy, hydroelectric power is dispatchable, and it accounts for 43% of the world's carbon-free electricity, followed by 28 and 22%, respectively, for nuclear and variable renewable energy.[6] The world's largest power

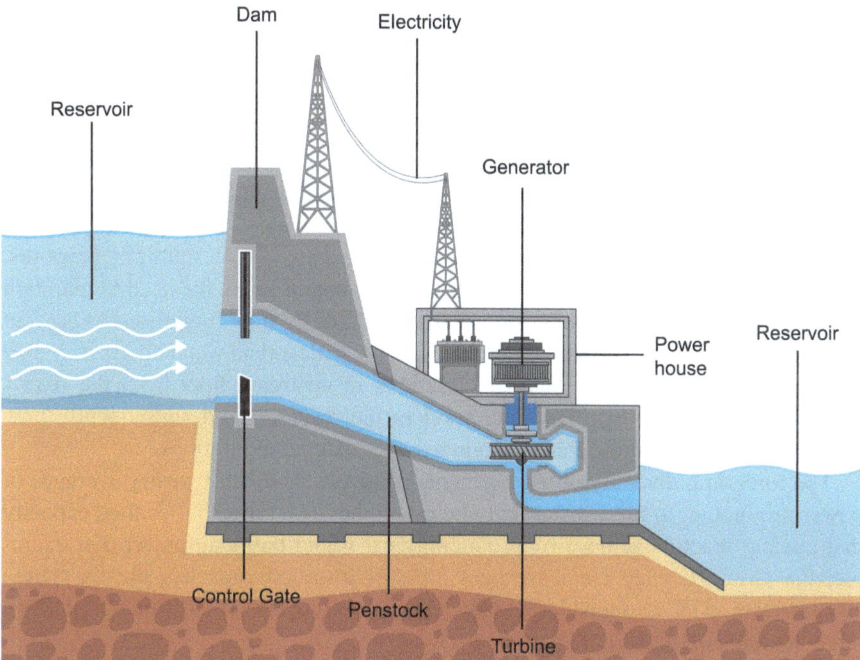


**Fig. 7.2** Components of a hydroelectric plant. (Source: iStock)

---

[6] https://www.iaea.org/bulletin/what-is-the-clean-energy-transition-and-how-does-nuclear-power-fit-in

plants are hydroelectric, led by China's Three Gorges Dam with an installed capacity of 22.5 GW. China also has the largest total hydroelectric capacity, with 341 GW, followed by the USA and Brazil with about 100 GW each, with Canada and Russia close behind. Some smaller countries may not have single hydroelectric plants or total outputs that rival these numbers, but their geography allows them to supply a large share of their power from hydroelectric plants. Norway leads the way with 99% of its power coming from hydroelectric plants.

The 6.8 GW Grand Coulee Dam on the Columbia River houses the largest hydroelectric generation in the USA, followed by the 2.7 GW Robert Moses plant at Niagara Falls. Dams are sometimes built-in series, such as the two highest dams in the USA, which are on the Colorado River. First to be built was the Hoover Dam completed in the 1930 with a capacity of 2.1 GW, and 200 miles upstream is the Glen Canyon Dam with 1.3 GW capacity.

The dams that hydroelectric plants require are expensive to build, but the plants incur no dollar cost for fuel, since their power comes from water stored in the dam's reservoir. Hence, using standard economic criteria, hydroelectric plants are ideal for providing reliable power, day and night. Setting the number and power of the water turbines—the installed capacity—for a hydroelectric power plant, however, depends on several variables. These include the annual flow of the river and the size of the dam reservoir. Turbine size is likely to be substantially larger than needed to convert the river's average flow to electricity. Since water turbines ramp from zero to full power in fractions of a minute, high powered turbines are indispensable for injecting large amounts of energy onto electric grids over for short periods of time to deal with extreme weather demands, sudden power plant failures, or other contingencies. Nevertheless, hydroelectric plants are also commonly used for load following and to provide baseload power while operating at less than installed capacity.

Coordinating hydroelectric power production with water flow requirements from large dams is often a balancing act, for such dams serve multiple purposes. Hydroelectric generation may be constrained, for example, by the need to maintain a flow adequate to provide downstream irrigation, or by flow limitations to prevent downstream flooding. The use of the reservoir for widespread water sports may also have impacts. Seasonal and annual variations in upstream flow from the river's watershed will also affect the water level in the reservoir and consequently how much hydroelectric power can be produced over extended durations.

The foregoing diversity of requirements causes hydroelectric yearly averages of power production to fall substantially below in the 90% range of installed capacity that nuclear reactors achieve by producing year-round baseload power. For example, the Grand Coulee Dam averages 35% of installed capacity, and the Three Gorges Dam, 51%. Nonetheless, larger dams, such as the Grand Coulee, produce large amounts of year-round baseload power.

Building large dams to house hydroelectric plants is extremely expensive, amounting to billions of today's dollars, and it may take 50 years or more to recover the capital costs of construction. For example, the 50-year loan from the US Treasury to build the Hoover Dam was paid off in 1987. Since it is nearly impossible to raise sufficient private capital to finance such massive, long-term projects, invariably they are built by government agencies backed by taxing power to cover the risks incurred.

Three federal agencies serve this purpose: The Army Corps of Engineers, the Bureau of Reclamation, and the Tennessee Valley Authority. All the dams in the USA with power of more than 0.4 GW are owned by these or other quasigovernmental agencies.[7]

Only about 7% of electricity in the USA comes from hydroelectric power, and over recent decades, that percentage has slowly decreased, mainly because there is a paucity of sites available for large dam construction. Even where there are locations that are suitable hydrologically for a large dam and reservoir, obtaining the rights to flood the land would likely bring severe social, economic, and environmental challenges. Such was the turmoil that arose in securing the land for the Grand Coulee dam reservoir in the 1940s. Sizable populations, including indigenous peoples, had to be moved and resettled. Farmers lost croplands that were difficult to replace, and the paths of spawning fish, such as salmon, were blocked. It was the federal government's leverage, arguing that the power was needed to win World War II, that finally overcame resistance to the dam.[8]

Nearly 2000 smaller dams dot the USA, with only a tiny fraction of them generating electricity. Most were built for irrigation, flood control, recreation, and other purposes. A fair number could be repurposed by incorporating hydroelectric generators. In locations where geographical or social constraint makes the creation of a reservoir impossible, run-of-river dams are often built, particularly on smaller but fast flowing rivers. In these, some of the river's flow is often diverted through a tunnel where a water turbine is located, and then returned to the river. One Department of Energy study concluded that 12 GW of hydroelectric capacity could be created where existing dams that presently have no hydroelectric facilities are located.[9]

***

Hydroelectric plants are not subject to the day-by-day intermittency of wind and solar energy. However, they are subject to longer term weather phenomena. The power of run-of-river dams varies with the water's seasonal flow, usually peaking during the spring months of snow melt and run off, with their nadir in the dryer months of summer and fall. Nevertheless, their variability would not be closely correlated with that of wind and solar energy generation and would likely improve the profile of residual demand curves. Reservoir water level management of larger dams normally allows hydroelectric plants to operate year-round largely unimpeded. However, in mountainous areas, spring may bring the danger of overfilled reservoirs as snowpack at higher elevations melts. Historically, high levels of precipitation have caused problems. In August 1983, for example, Hoover Dam overflowed through its two spillways, which fortunately had sufficient capacity to prevent the dam from overtopping. Multiyear droughts that appear to be getting worse with global warming can have disastrous results. The current situation in the southwestern USA illustrates the point.

---

[7] http://www.idahopower.com/AboutUs/OurPowerPlants/Hydroelectric/hellsComplex.cfm

[8] https://www.washingtonhistory.org/wp-content/uploads/2020/04/water-rises.pdf

[9] https://www.energy.gov/eere/water/articles/hydronext-fact-sheet

Lake Mead and Lake Powel behind Hoover and Glen Canyon dams are two of the largest reservoirs in the country. Filling these reservoirs with Colorado River water took 3 years following the construction of each dam. They each cover many square miles, and their extensive shorelines have become the center for water sports in the southwest. But now, the multiyear drought coupled with increased evaporation from rising temperatures is causing reservoir levels to drop precipitously. Lake Mead has lost 70% of its volume, lake Powell 77% of its. The water level in lake Mead has dropped 131 feet in 4 years, causing the power that Hoover Dam can produce to be cut in half, and if it drops only 32 feet more, the water turbines will cease to operate.[10] The numbers for lake Powell are similarly grave.

While this loss of power from the two dams would seriously impede progress toward the nation becoming carbon neutral, other existing fossil fuel plants could pick up the load. But if the water level in lake Mead falls to its dead-pool level at which no water can get through the dam, it would be catastrophic. 40 million people in seven states and a part of Mexico depend on the Colorado River for drinking water, and it irrigates more than 8000 square miles of agricultural land. To somewhat alleviate the problem, President Biden negotiated an agreement between the states to reduce their water consumption by 13%. Whether that will be sufficient is an open question.

Multiyear droughts elsewhere have impeded hydroelectric power as well.[11] Canada is the third largest producer of hydroelectricity in the world and counts on its widespread system of dams to provide 60% of its electricity. However, a coast-to-coast drought is curtailing hydroelectric power production. Quebec is Canada's largest generator of hydroelectric power, with plans to export copious quantities of electricity to the northeastern US states of Vermont, Massachusetts, and New York. Weak snow cover and a paucity of rain have reduced the flow of water into its reservoirs to an extent that power exports are being reduced.[12] British Columbia, normally Canada's second largest producer of hydroelectricity, has been experiencing drought since mid 2022, with large reservoir levels lower than they have been in many years.[13] Thus, the impacts of climate change are affecting even our most trusted sources of renewable energy.

***

The world's oceans present a colossal source of energy in their tides, currents, waves, and thermal gradients. With one exception, however, capturing energy on a significant scale has encountered economic, technical, and environmental problems that have yet to be overcome. That one exception is tidal energy. While there has

---

[10] https://www.doi.gov/pressreleases/interior-department-announces-actions-protect-colorado-river-system-sets-2023

[11] https://www.wsj.com/us-news/climate-environment/canada-had-designs-on-being-a-hydro-superpower-now-its-rivers-and-lakes-are-drying-up

[12] https://www.hydroreview.com/business-finance/business/unfavorable-weather-conditions-lead-hydro-quebec-to-reduce-its-exports/#gref

[13] https://www.bnnbloomberg.ca/drought-in-western-canada-impacting-hydropower-production-as-reservoirs-run-low

been no significant use of tidal power in the USA thus far, large projects in France, Korea, and elsewhere have created electric generation capacity measured in the hundreds of megawatts.

Tidal energy is renewable in that it arises from the gravitational interaction between the Earth and moon. Hurricanes and other large storms affect the tides. They bring low atmospheric pressures which cause sea level rise and therefore the tides to be higher. Otherwise, tides are regular and completely predictable. The tides' characteristics do not vary with the season or day-to-day weather variations. All coastal areas experience two high and two low tides every 24 h and 50 min.

Large tidal projects are located on estuaries or other bays or inlets where the tidal swells are large, and the resulting currents strong. A tidal dam called a barrage is built to wall off the estuary from the ocean. It only needs to be taller than the highest tide at that location. However, it may be quite long to span the mouth of the reservoir that it creates. Figure 7.3 shows the barrage of the La Rance Tidal Power Station in Brittany, France, until recently the world's most powerful tidal plant. Turbines located in the barrage's base generate electricity when water from the reservoir passes through them in the outward direction. For each 12-h tidal cycle, the barrage sluice gates first allow water to fill the reservoir as the tide rises. The gates then close trapping the water in the reservoir. As the tide ebbs, water flows outward through turbines, generating electricity until the water levels of either side of the barrage equalize.

Note that electricity is produced for only 6 of the 12 h of each tidal cycle. Thus, just as in the case of solar and wind generation, there is a residual demand that must be met by other means. However, the effect that tidal power will have on renewables' residual demand is known precisely in advance, since it doesn't depend significantly on the weather.

**Fig. 7.3** The La Rance Tidal Power Station. (Source: Wikimedia Commons)

The obstacles to building barrages to expand the use of tidal power substantially are similar to those confronting the building of large river dams to increase hydro-electric capacity. They include the paucity of geographically suitable sites, exceeding large capital costs, loss of recreational seashore, and environmental damage to wetland or marine life. There are smaller scale alternatives that don't require a barrage. These include arrays of small water turbines placed on the seabed that capture energy tidal inward and/or outward currents, instead of differences in water level of either side of a barrage. They, however, are not likely to generate nearly as much power as do barrage-based systems.

***

Geothermal energy is renewable, and it is brought to the Earth's surface by water. The energy originates deep within our planet's core, coming from heat emanating from the radioactive decay of thorium, uranium, and polonium. Active geothermal locations are found close to tectonic plate boundaries near where most volcanic activity occurs. The ring of fire around the Pacific Ocean, which includes the western USA, consists of a series of such plate boundaries. Near the boundaries, hot magma comes closer to the surface than in other parts of the world, and where subsurface reservoirs exist, their water is heated to hundreds of degrees Fahrenheit. It does not boil, however, since pressure increases with depth and the boiling temperature along with it. These reservoirs may be two or more miles beneath the surface. In a few localities, permeable rock allows breaks through the Earth's crust to create geysers sending columns of steam and water into the air.

Old Faithful is the most famous of the more than 500 geysers in Yellowstone National Park. These geysers, however, are not allowed to be disturbed and therefore cannot produce usable energy. The only other place where geysers exist in the USA is at The Geysers Field in Northern California, where energy can be diverted legally for power production. The Geysers sit atop a large magma chamber more than 4 miles beneath the surface. The field supports 18 geothermal power plants producing more than 800 megawatts of electricity, equivalent to 20% of California's renewable energy.[14]

Three types of plants produce electricity at Geysers Field. Dry steam plants bring superheated steam to the surface from geothermal reservoirs to power steam turbines directly. Flash steam plants take high-pressure hot water from deep inside the Earth and decrease its pressure in flash tanks to where it boils, thus converting it to steam that drives turbines. Most geothermal power plants are the flash steam variety. If the water reaching the surface has a temperature of less than 360 °F, flashing it would produce steam too cool to drive a turbine. Instead, its heat is employed in a binary cycle. In it, the heat from geothermal hot water is transferred to another working fluid, usually an organic compound that has a lower boiling point than water. The organic fluid vaporizes and drives a turbine. For all three plant types, the steam or other working fluid exits the turbine, is condensed, and is then recycled back into the system for repeated use.

---

[14] https://en.wikipedia.org/wiki/The_Geysers

Of domestic locations, only The Geysers field includes dry steam wells. Across a number of western states where the geology is amenable to geothermal energy, some 30 plants are in operation, the flash steam plants being most numerous. They supply 3.7 GW of geothermal electricity. The number of projects proposed or underway could increase that number by several GW.[15] The USA leads the world in geothermal power production. Indonesia and the Philippines follow,[16] taking advantage of their positions on the "Ring of Fire" that offer sites geographically conducive to geothermal capacity.

Thus far, geothermal energy plants operate only atop locations where hydrothermal reservoirs satisfy three key elements: heat, fluid availability, and permeability, the latter being where water can move freely through deep underground rock. Geothermal could play a much larger role in fighting climate change if enhanced geothermal energy methods could unlock the heat from deep in the rock that underlies most land masses. That rock is hot, but it's not sufficiently permeable for water to flow through it, even if there were any there.

An enhanced geothermal system would require drilling 2–4 miles into the rock and injecting a fluid. Fluid pressure must then cause pre-existing factures to open, such that the enhanced permeability of the now-fractured rock would allow fluid to circulate, absorbing heat from a large volume of the rock. The resulting artificial hydrothermal reservoir would then serve the same purpose as the natural reservoirs in present-day geothermal energy plants: Fig. 7.4 shows the two types of wells required, an injection well, and one or more production wells to return the heated water back to the surface. The fractured rock lies between the two wells. If the water coming from the production wells is not hot enough for its steam to drive a turbine directly, then the binary cycle shown in Fig. 7.4 would be employed. In it, a heat exchanger transfers heat from the water to an organic fluid with a lower boiling point. The water is then pumped into the injection well. The organic fluid circulates through the turbine to generate electricity and then is condensed to liquid form pumped to the heat exchanger.

Research on enhanced geothermal systems is making rapid strides, but the task is formidable. The well system shown in Fig. 7.4, however, has similarities to hydraulic fracturing (i.e., fracking) for oil and gas, and experience gained there should be valuable.[17] Some risks are also analogous to those of fracking. Success is possible only in areas of the county where suitable rock beds exist, and even there, the risk is significant of drilling miles-deep dry holes, in which no suitable bed of fractured rock exists. Like fracking, some enhanced geothermal projects may also result in earthquakes of unacceptable magnitude.

---

[15] https://www.energy.gov/eere/articles/nrel-2021-us-geothermal-market-report-released

[16] https://www.thinkgeoenergy.com/thinkgeoenergys-top-10-geothermal-countries-2022-power-generation-capacity-mw/

[17] https://www.wsj.com/business/energy-oil/frackers-are-now-drilling-for-clean-power-fb613dbf?mod=djemclimate

**Fig. 7.4** Enhanced geothermal electricity generation. (Source: iStock)

A yet more ambitious geothermal undertaking is a closed loop system,[18] some-times referred to as an Eavor-Loop™. It entails drilling two wells to depths of 2 or more miles beneath the surface. Deeper than where hydro reservoirs or fractured rock may lie, the well tips must penetrate the solid rock that underlies nearly the entire continent. The system layout is similar to the two-well system shown in Fig. 7.4. Now, however, the well tips are in solid rather than fractured rock, and the injection and production wells are connected by a substantial number of pipes running between them. Thus, a closed loop system is created. Natural convection causes a fluid, which doesn't need to be water, to circulate through the pipes and wells, picking up heat from the dry hot rock and delivering it to the heat exchanger. The unique advantage of such closed loop system is that they could be located anywhere on the continent, since they don't depend on an underlying reservoir of permeable rock.

Enhanced open or closed loop geothermal systems may offer a solution to the global warming dilemma. Geothermal energy is renewable, and like nuclear energy, it would produce power that is free of greenhouse gas emissions. Geothermal plants may also have the capability for load following to match the time variation in

---

[18] https://www.eavor.com/wp-content/uploads/2021/08/Science-Direct-Closed-loop-geothermal--energy-recovery-from-deep-high-enthalpy-systems.pdf

electricity demand,[19] while requiring little land relative to other forms of renewable energy. But there is much to do before geothermal can supply a substantial fraction of energy demand. Research and development are progressing, but they need to be accelerated, for more than any other form of renewable energy, over the long term, advanced geothermal systems have the potential to play a leading role in curbing global warming.

---

[19] Ricks, W., et al. (2022) "The Value of In-Reservoir Energy Storage for Flexible Dispatch of Geothermal Energy," *Applied Energy* 313, 118807.

# Chapter 8
# To Store and Convert

**Keywords** Electricity storage · Roundtrip efficiency · Battery storage · Pump storage · Compressed air storage · Gravitational storage · Thermal storage · Hydrogen and ammonia storage

When I think of a lithium-ion battery, it's one of the those in my cell phone or laptop that first come to mind, not those that make up massive storage banks such as shown in Fig. 8.1, still under construction at Moss Landing, California. At present, it is the largest battery storage facility in the USA. The amount of energy stored in several such large energy banks coming into existence indeed is impressive. However, confining huge quantities of energy—whether it be lithium in batteries, gas in tanks, or water behind dams—also presents ever-present risks.

On January 16, 2025, such dangers became all too apparent. The Moss Landing battery bank caught fire, and the bank's batteries burned for days.[1] The release of

**Fig. 8.1** Moss Landing battery storage project. (Source: Bloomberg)

---

[1] https://www.technologyreview.com/2025/02/13/1111843/battery-fire-moss-landing-power-plant/

toxic heavy metals, including nickel, manganese, and cobalt, prompted the evacuation of more than 1000 residents and contaminated the air, the surrounding soil, and the waters of the Monterey estuary.[2] The study of environmental and health impacts is ongoing.

The Moss River battery bank, when completed, will produce 750 megawatts (MW) for 4 h.[3] With the risk of fire adequately controlled, it will provide California's electric grid with the power to compensate for the diurnal intermittency of the growing capacity of wind and solar energy. Headline publicity associated with battery storage, however, is often exaggerated or misleading. For example, "U.S. power grid added battery equivalent of 20 nuclear reactors in past 4 years."[4] That 20 refers to 20 Gigawatts (GW) of power that the battery banks can deliver. With the assumption that the reactors are of the common size of 1 GW, a simple calculation shows it would take 27 facilities of the final size of the Moss Landing to provide the electricity of 20 reactors for 4 h. It would take more than 80 such battery facilities to get through the night. As we shall see, battery banks provide useful fill in for time periods of a few hours or less, but for durations of a day or more to compensate for overcast skies or wind lulls, they do not constitute a viable solution.

Battery banks are particularly useful for power regulation, which is the control of random fluctuations and disturbances in supply and demand on time scales ranging from thousandths of a second to minutes. Such disturbances degrade the quality of electricity on the grid, resulting in frequency irregularities and voltage sag, spikes, and flutter.[5] Regulation is becoming more challenging as turbine-based nuclear, hydroelectric, and fossil fueled plants are replaced by variable renewable energy (VRE), since the massive synchronous inertia of spinning turbines has been the principal source of grid stability. Batteries providing rapid injections of power that are needed for only a few seconds to control such phenomena. Although not widely used at present, flywheels and super capacitors may serve some of the same purposes.

The power and energy storage capacity of large banks of batteries will become useful for injecting power into electric grids for longer times, ranging between a few minutes and a few hours. Increasingly, they serve to compensate for some of the intermittency of solar and wind energy, for example by countering sudden changes in renewable energy supply that result from wind gusts or fluctuations in cloud cover. Moreover, when solar farms account for even 30 or 40% of electricity, the farms are likely to supply more electricity during the midday hours than there is demand. Rather than curtailing the solar energy, the excess may be used to charge the batteries. They may then be discharged at dusk to augment the electricity supply

---

[2] https://www.nytimes.com/2025/02/10/us/california-battery-plant-fire.html

[3] https://investor.vistracorp.com/2022-01-24-Vistra-Announces-Expansion-of-Worlds-Largest-Battery-Energy-Storage-Facility?utm_source=chatgpt.com

[4] https://www.theguardian.com/environment/2024/oct/24/power-grid-battery-capacity-growth

[5] Denholm, P. et al., (2010). The Role of Energy Storage with Renewable Electricity Generation NREL/TP-6A2-47187.

from solar farms, which diminishes rapidly as the sun sets. By doing so, they reduce demand for ramp rates in power that are too rapid for thermal power plants to follow.

Battery storage facilities are also proving valuable in dealing with emergency situations. Indeed, they were critical in avoiding rolling blackouts in Texas from the record setting heat of September 26, 2023.[6] Although solar farms were supplying substantially less than half of the electricity, when the wind died and solar output plummeted with the setting sun, rolling blackouts were imminent. However, quick insertion of power from battery storage bolstered the grid for an hour or more until demand subsided and slower acting gas-fired plants increased in power and restore stability. However, that short-term power input totally discharged the battery banks throughout Texas. Yet, the electricity they supplied amounted to less than 3% of the power demand during the critical time period.[7]

***

Electricity circulating through the wires of a grid cannot be stored as electricity. Therefore, when we speak of electricity storage, we are actually describing energy being stored in some other form, such as chemical, mechanical, or thermal energy. The batteries thus far discussed store it as chemical energy. With decreasing costs and improved performance, lithium-ion batteries are garnering most attention. Other kinds of batteries may offer different combinations of desirable properties: economy, power, longevity, and sustainability. Some are quite familiar, such as lead acetate batteries found in our cars. Feasibilities of iron-air batteries,[8] as well as sodium-ion nickel-hydrogen and other material combinations are also being investigated. Flow batteries based on zinc compounds are still in the research and development stage.[9] In them, the electrolyte flows between connected tanks, with the flow direction being reversed when the cell is charging from when it is discharging. These may eventually offer greater improvements in power and discharge hours over what is possible with traditional battery compounds and configurations.

While batteries' chemical storage of electricity is valued for a the afore stated purposes, they have their limitations. Even with foreseeable improvements, battery banks such as at Moss Landing will most likely be limited to powers of less than a Gigawatt with discharge times of substantially less than a week.[10] Their affordability becomes questionable for storing the huge amounts of energy needed to fill in when VRE becomes a substantial fraction of electricity generation.[11] Lithium-ion batteries have a lifespan of up to 15 years.[12] Thus, the capital cost of replacing facilities such as Moss Landing is incurred far more frequently than for electricity

---

[6] https://www.wsj.com/business/energy-oil/giant-batteries-helped-the-u-s-power-grid-eke-through-summer-a68425fd

[7] Ibid.

[8] https://formenergy.com/technology/battery-technology/

[9] https://www.battery.associates/post/state-of-art-of-flow-batteries-a-brief-overview

[10] Ibrahim, H. et al. (2008), "Energy storage systems—Characteristics and comparisons," *Renewable and Sustainable Energy Reviews*, 12, 5.

[11] https://www.technologyreview.com/2018/07/27/141282/the-25-trillion-reason-we-cant-rely-on-batteries-to-clean-up-the-grid/

[12] https://www.renogy.com/blog/how-long-do-lithium-batteries-last

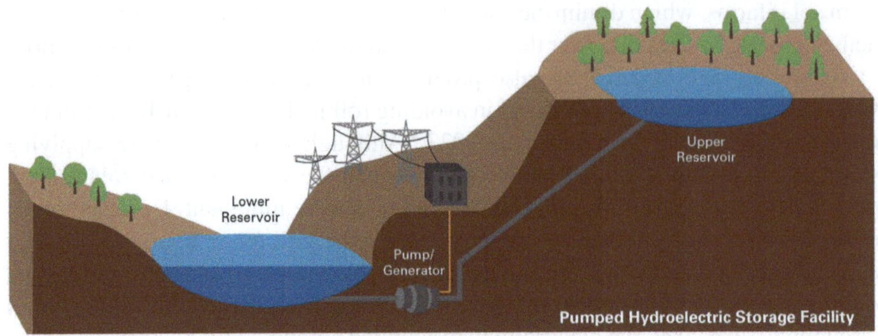


**Fig. 8.2** Pumped hydroelectric storage plant. (Source: Dominion Energy)

generators: 25–30 years for VRE and gas-fired plants, 60–80 years for nuclear reactors. Worse yet, lithium-ion batteries tend to discharge when they lie idle for several weeks. For powers in the GW range, discharge times of a week or longer, or storage for weeks or longer, different forms of energy storage are needed. At present, foremost among these is pumped hydroelectric storage.

Pump storage facilities are closely related to dam-based hydroelectric power plants. They consist of two reservoirs at different elevations such as shown schematically in Fig. 8.2. A tunnel connects the two reservoirs, and reversible water turbines are located at the inlet to the lower reservoir. Except where solar energy is prominent, such systems shift electricity production from night when demand is low to daytime when it is high.[13] During the night, the grid's other generators can produce more power than needed. The excess is used to pump water from the lower to the upper reservoir by running the water turbines in reverse. Then, during the day when more power is demanded, the turbines run in the forward direction, generating electricity from the force of the water as the upper reservoir empties into the lower one. Since the price of electricity is lower at night when the grid can produce more electricity than is demanded, and higher during the day when demand is higher, income is derived from the price difference, allowing the cost of building the pump storage plant to be recovered.

At present, the Bath County Virginia pump storage is the largest in the USA.[14] The dam for its upper reservoir is 460 feet high, and for its lower reservoir 135 feet high. The turbines generate 3 GW of power coming from an elevation difference between reservoirs of 325 ft. In the USA, only the Grand Coulee Dam generates more hydroelectric power. Each of the Bath County reservoirs stores the equivalent of 24 GWh of electricity. A comparison with the Moss Landing battery system when it is completed is shown in Table 8.1. The power generated in the Moss Landing plant is 2.5 times as much as that of the pumped hydro facility. However, the discharge at full power for the pump storage plant is twice as long as for the battery bank.

---

[13] http://www.virginiaplaces.org/energy/pumpedstorage.html

[14] http://www.virginiaplaces.org/energy/pumpedstorage.html

**Table 8.1** Energy storage characteristics

|                            | Power (MW) | Discharge time (h) |
| -------------------------- | ---------- | ------------------ |
| Moss landing battery       | 7500       | 4                  |
| Bath county-pumped hydro   | 3000       | 8                  |

Often, storage facility operators find it more economical to charge and discharge at partial power for longer periods of time. For example, the pumped hydro plant might go on a 24-h cycle charging and discharging at 2 GW for 12 h, instead of at 3 GW for 8 h. In either situation, the electricity is stored when excess supply is available, and prices are low, and discharged when demand and thus prices are high, most frequently on a 24-h cycle.

There are nearly 50 pump storage systems in the USA, and they provide more than 95% of the utility scale energy storage.[15] As at Bath County, they are most frequently used for more nearly leveling day and night electrical generation, thus allowing a larger fraction of the electricity generated to be 24/7 baseline power, which can be provided by the lower fuel cost of nuclear power plants. Like the hydroelectric plants discussed in Chap. 8, pump storage water turbines can start and reach full power in 5–10 min. Hence, if a sufficient volume of water is retained in the upper reservoir, they can be used to fill in for missing power when solar power plummets at the end of the day, or to provide needed electricity if a large generator must be shut down; for example, the Bath County facility could replace a GW nuclear reactor for 24 h.

There are several pump storage projects under way in the US West, but further work will be increasingly constrained by the lack of suitable geological sites for expansion. In theory, very large plants could be built by having the upper reservoir being a large lake, and the lower reservoir deep below in an abandoned mine, salt dome or other cavern, like what is used for storing natural gas. But such ideas are only in the early stages of exploration.[16]

Where solar energy is abundant the day–night pump storage cycle is often the reversed from that of the Bath County plant described above. In Spain, for example, hydroelectric and nuclear plants along with pump storage provide the electricity during the nighttime hours. During the day when solar generation exceeds demand, the hydroelectric plants are shut down, the nuclear plants reduced to minimum power, and the pump storage reservoirs are filled, powered by the excess solar energy.

***

In addition to batteries and pumped storage, other forms of energy storage are under development. Among them is compressed air storage, although to date, only a few compressed air systems have been built. Strictly speaking, these are not pure storage systems.[17] Rather they utilize storage as a way of greatly increasing the efficiency of natural gas fueled turbines. In a standard gas turbine, as much as

---

[15] https://www.energy.gov/eere/water/pumped-storage-hydropower

[16] Lyu, X, et al. (2022). "Pumped Storage Hydropower in Abandoned Mine Shafts: Key Concerns and Research Directions," *Sustainability* 14, 16012.

[17] https://www.azom.com/article.aspx?ArticleID=22859

two-third of the electricity produced is consumed internally in driving its compressor. Hence, for use in conjunction with compressed air storage, the compressor is eliminated from the turbine. Instead, atmospheric air is compressed to greatly increase its pressure. The high-pressure air passes down through boreholes to a leak tight cavern where it is stored. To produce power, the compressed air passes up through a second set of boreholes into the gas turbine where it mixes with methane and undergoes combustion. Without the need for a compressor, the turbine produces about three times as much electricity as it otherwise would.

The forgoing system can avoid the use of fossil fuel if the gas turbine if fueled with hydrogen produced by electrolysis, employing electricity from renewable or nuclear sources. An alternative that requires no fuel other than the electricity used to compress the air has been termed advanced compressed air storage.[18] Advanced systems take advantage of the massive amount of heat produced in the process of compressing air. Instead of letting it dissipate as waste heat, it heats a fluid which is stored in a well-insulated tank. When power is needed, that heat is inserted into the compressed air streaming from the reservoir to drive a turbine generator, which requires no additional fuel. Two such systems are operating in China, and one in Canada.[19]

The power of a compressed air system is proportional to the sizing of the air compressor and the turbine, while the energy stored—or discharge time—depends on the size of the cavern and the pressure of the stored air. Securing suitable underground leak-proof caverns large enough to store sufficient compressed air is a challenge. Salt domes would seem a likely choice. However, they are ineffective because they are at constant volume. Thus, as air is supplied to the turbine, the pressure in the dome decreases as does the electric power produced by the turbine.

Porous sandstone aquifers are preferable.[20] In them, the stored air pressure stays constant, no matter whether the reservoir is full or nearly empty. Constant pressure is maintained because air pumped into a sandstone aquifer forms a large bubble, displacing the water around it. As air is released to the turbine, the size of the bubble decreases, but its pressure does not, and the power output of the turbine also stays constant. Unfortunately, constant pressure sandstone aquifers are also needed and widely used for natural gas storage. Accordingly, competition to secure storage space for compressed air will be intense. An innovative alternative under development stores the compressed air underwater in inflatable balloon like structures.[21] Thus, the water pressure keeps the compressed air at constant pressure, since as air is extracted, the balloon's volume contracts.

Gravity can also serve as a basis for storing energy. In such a system, electric motors would lift heavy weights to a substantial height. Then when energy is needed

---

[18] https://royalsociety.org/-/media/policy/projects/large-scale-electricity-storage/large-scale-electricity-storage-report.pdf

[19] Ibid.

[20] https://www.osti.gov/servlets/purl/6270908

[21] https://electricalindustry.ca/changing-scenes/1785-world-s-first-utility-scale-underwater-compressed-air-energy-storage-system-activated-in-lake-ontario/economis

the weights would be dropped, causing the motors to become generators by running in reverse. Very heavy weights and large heights would be required. A project in the United Kingdom plans to use old coal mine shafts.[22] Weights totaling 12,000 tons will be hoisted from the bottom of each half mile-deep mine shaft, using surplus electricity. The weights will then be dropped when power is in short supply. Friction and other losses would reduce the energy output to less than the theoretical limit of 20 MWh per drop. Controlling the drop speed would allow the system power and discharge time to be specified. For example, the theoretical limit for a 15-min drop would be 80 MW, and 40 MW for a 30-min drop.

***

Hydrogen is a widely used commodity, an essential feedstock for many industrial processes. If produced without creating greenhouse gases, it becomes a valuable ingredient in decarbonizing large parts of the transportation and industrial sectors of the economy. At present, nearly all hydrogen is produced by reforming natural gas, a process in which natural gas is combined with high-temperature steam in a catalytic converter to produce hydrogen and the unfortunate $CO_2$ byproduct. Decarbonizing this process would include the expense of employing carbon capture described in Chap. 10. Long-term research is under way on possible alternative for producing hydrogen from methane without generating $CO_2$. It would use a high-temperature pyrolysis process to split methane into hydrogen and graphite, i.e., solid carbon.[23]

In nearer term, an attractive alternative exists for producing hydrogen without generating greenhouse gases. Electrolysis of water, using electricity from nuclear or variable renewable energy power plants,[24] yields only hydrogen and oxygen. In effect, the carbon-free electricity is stored as hydrogen. Pairing a nuclear reactor with wind or solar energy to simultaneously produce both electricity and hydrogen may also be effective in dealing with the VRE intermittency dilemma detailed in Chap. 5: The reactor operates continuously at full power. It supplies the erratic residual demand caused by intermittent VRE generation, and its remaining output is sent to the electrolyzer to produce hydrogen. The hydrogen production will also be intermittent but controlled to complement the residual demand requirement such that the sum of electricity produced to supply residual demand and that supplied for hydrogen production will equal the reactor's full power, allowing it to operate steady state, 24/7.

Once produced, hydrogen can be stored in high-pressure tanks for extended periods of time. It may then be converted back to electricity in two ways. The first is to employ fuel cells, which run the electrolysis process backward, combining hydrogen and oxygen to produce water and electricity. The second is to use it to fuel future gas turbines, designed to burn hydrogen instead of methane.

---

[22] https://www.theguardian.com/environment/2019/oct/21/how-uks-disused-mine-shafts-plan-to-store-renewable-energy

[23] https://www.third-derivative.org/blog/hydrogen-produced-from-methane-pyrolysis

[24] https://www.nrel.gov/docs/fy04osti/36705.pdf

Fuel cells are compact and envisioned to decarbonize much of the transportation sector. For automobiles, however, battery power seems more likely to dominate the market for at least two reasons. Both batteries and fuel cells must be recharged. Much progress has already been made in implementing a national network of battery charging stations, while few if any hydrogen charging stations exist. Moreover, setting up charging stations for fuel cells is more difficult than for batteries. While only adequate electrical connections are needed for battery charging, fuel cell charging requires large high-pressure storage tanks on site or a piping network to supply the hydrogen.

While batteries are most likely to dominate the electrification of automobiles, they are compact but heavy, and their weight plays a large role in determining how far a car can travel between recharging. Conversely, a tank of hydrogen of the size and pressure needed to power an electric car with a fuel cell would weigh much less than a battery, but its volume would be much larger, likely far too large to fit within the footprint of most automobiles. Fuel cells are more appropriate for larger long-haul vehicles, for which tank size is less of a limitation and where needed hydrogen storage and access can be placed hundreds of miles apart or located at the beginning and end of frequently repeated trips. Interstate trucking and the railroads could be more effectively electrified using hydrogen fuel cells than batteries.

Present day gas turbines use methane to generate electricity. However, research is ongoing to increase the fraction of hydrogen that these turbines can burn, with the objective of eventually creating turbines fueled completely by hydrogen. However, since both fuel cells and gas turbines have efficiencies of about 60%, and electrolyzer efficiencies are roughly 70%,[25] the round-trip electricity $\rightarrow$ hydrogen $\rightarrow$ electricity efficiency is only about 40%. Both batteries and pump storage are more efficient than such round-trip hydrogen storage methods. More efficient uses of decarbonized hydrogen would be to employ it as feedstock for the numerous industrial processes for which it is required. Hydrogen can replace fossil fuels in supplying the high-temperature heat for industrial processes that account for nearly a quarter of greenhouse gas emissions.[26] Steel and cement are the industries emitting the greatest amounts of greenhouse gases. Taken together, they emit 15% of the $CO_2$ worldwide.[27] In steel production, hydrogen may take the place of coal. Presently, coal heats blast furnaces to temperatures of more than 1400 °C (2600 °F) to reduce iron ore to elemental iron. Instead, hydrogen also react with iron ore at high temperatures in a three-step combustion process that has the net result of producing elemental iron and water. Hence, no $CO_2$ is emitted. After adding scrap iron and alloying materials to make steel, the process may be completed using electric arc heating.

In fact, another approach is used increasingly to make steel without producing $CO_2$. In the mature industrial societies of Europe, the USA, and elsewhere—and to

---

[25] https://www.energy.gov/fecm/how-gas-turbine-power-plants-work

[26] See Fig. 2.1.

[27] https://cleantechnica.com/2020/09/14/reducing-emissions-from-cement-and-steel-production/

some extent in China—enough scrap metal is recycled from demolished bridges, buildings, vehicles, and machinery of all sorts to satisfy a substantial percentage of the demand for new steel. Smelting iron ore isn't required to produce steel from scrap iron; arc furnaces, which can be powered by carbon-free electric generation, can accomplish the task. Indeed, increasing numbers coal-free steel mills are coming into use in several countries.

Worldwide, however, greater emissions of carbon dioxide come from the kilns producing cement than from making steel. Hydrogen could replace fossil fuels in heating the kilns. However, the chemical reaction that creates cement also produces $CO_2$, and to deal with it, carbon capture and sequestration, discussed in Chap. 10, would be required.

Challenges will be encountered expanding hydrogen's usage. They will abound in developing the infrastructure for storing and transporting a gas that is lighter than natural gas. Hydrogen's light weight causes it to leak more easily from tanks and pipes. Moreover, leaks at an early stage are not easily detected since a hydrogen flame is invisible. It also causes embrittlement and consequent weakening of metal piping and other fixtures. Hence, existing pipeline networks built to distribute natural gas are unlikely to be safe for hydrogen transport. Likewise, redesigning gas turbines or other equipment powered by natural gas to be fueled by hydrogen seems a doable but arduous process.

Using hydrogen to make ammonia with the Haber-Bosh process may overcome some of these difficulties. While hydrogen is difficult to transport and store, ammonia changes from gas to liquid at room temperature when compressed to less than eight atmospheres pressure. It then stores a great deal more energy per cubic foot than gaseous hydrogen. Moreover, because millions of tons of ammonia are produced annually to make fertilizer, there already exist a functioning network for the transport and storage of ammonia. Ammonia can thus be a hydrogen carrier,[28] which undergoes the Haber–Bosch process in reverse—called cracking—to release the hydrogen for use.

Longer-term research is ongoing to develop gas turbines fueled by ammonia, and possibly even ammonia-based fuels to decarbonize container ship and aircraft fuels. There are, however, also challenges in dealing with wide-scale uses of ammonia. It is both flammable and toxic. Enforceable safety measures must accompany its use. Because it is widely considered only as a harmless fertilizer, numerous fatal explosions have resulted from its careless handling and storage. That would need to change.

***

Thermal energy—i.e., heat—storage is different from the battery, pumped hydro and the other storage methods that are powered by electricity. With heat storage, there need be no electricity input, for nuclear, geothermal, or concentrated solar

---

[28] https://kleinmanenergy.upenn.edu/wp-content/uploads/2023/03/KCEP-Digest53-Ammonias-Role-Net-Zero-Hydrogen-Economy.pdf

**Fig. 8.3** Concentrated solar energy plant. (Source: NREL)

energy can produce carbon free heat directly. Stored heat may be used for space heating, desalinization, and if its temperature is high enough to power a large number of chemical processes. In the power industry, the most experience in high-temperature heat storage has been with molten salts, and that has come primarily from large concentrated solar energy power plants, such as the one in Nevada.

The Crescent Dunes plant on the Nevada desert, pictured in Fig. 8.3, is representative. A large array for mirrors focused on a single point or line concentrates the solar energy. Crescent Dunes is a tower configuration in which 10,000 billboard sized mirrors concentrate sunlight on the top of a 640 feet tower.[29] Molten salt circulates through pipes to the top of the tower, where the solar radiation heats it to more than 540 °C (1000 °F). The molten salt then descends to ground level. In earlier systems, the fluid was immediately circulated through a heat exchanger, transferring heat to generate high-pressure steam which drove a turbine generator to produce electricity. Thus as with solar panels, electricity was only generated while the sun was shining.

An innovation of the Crescent Dunes plant is not to create steam immediately but to store the solar energy as heated molten salt in a very large and well insulated tank. The energy is saved in what is called the hot tank until power is needed at night or on a cloudy day. At that time, the molten salt circulates from the hot tank through a heat exchanger to create steam to power the turbine-generator and then to a cold tank that is similar in design to the hot tank. When the sun is shining, the molten salt circulates through the tower to be heated and repeat the cycle.

Even with the gains from this innovation, concentrated solar energy has not seen the growth that solar panels have for several reasons. First among them, the cost of solar panels has plummeted over the last decade making their use less expensive

---

[29] https://insideclimatenews.org/news/16012018/csp-concentrated-solar-molten-salt-storage-24-hour-renewable-energy-crescent-dunes-nevada/

than building a concentrated solar system. In addition, solar panels can be spread out over several different areas, making construction more straight forward than for the mirrors that must be precisely placed together in a specified geometry on level terrain. Hence, while the future for concentrated solar energy seems uncertain, thermal storage using molten salt is now well tested and can be used to deal with the problem of integrating nuclear energy with VRE on electric grids.

In subtropical regions where year-round solar energy is plentiful, it has been proposed that nuclear reactors could be located next to concentrated solar energy plants. Since steam derived from the molten salt can be hotter than the temperature of steam from the reactor, the two could be blended to increase efficiency, while the reactor would also serve as supplementary nighttime energy and as backup for cloudy days.[30] Thermal storage in molten salt can also be employed in conjunction with advanced reactors, as detailed in Chap. 14, to provide flexible power to compensate for the intermittence of wind and solar energy.

***

Energy storage systems will find increased use buying and selling energy on minute-by-minute or hour-by-hour basis to smooth peaks and crevasses in residual demand profiles, and on 24-h storage cycles to shift electricity supply between day and night to meet demand more economically than otherwise would be the case. In such situations, the difference between buy and sell prices must be great enough to compensate for the less than perfect turnaround efficiency of the storage systems. For example, if round trip efficiency is only 50%, the electricity must be sold at least double the price at which it was purchased just to break even.

Energy storage operating on a 24-h cycle may be viable, but providing storage for longer durations becomes increasingly fraught. On a grid heavily dependent on solar energy, for example, seasonal shifting of buying energy in the summer and selling it in the winter yields only a single buy-sell cycle per year. Capital costs—including finance charges—must be paid by selling power over the life of the storage facility. An annual buy-sell cycle would necessitate either a prohibitively high energy price or exorbitant monthly storage fees.

It may be tempting to also view energy storage as a means of dealing with the challenges presented by extended periods of overcast skies or calm winds. Solar output would be reduced to as little as 10% of installed capacity; wind turbines may produce no energy at all. Making matters worse, such conditions are likely to be exacerbated by extremes of heat or cold resulting in record electricity demand. To serve as backup, large amounts of energy in MWh would need to be stored.[31] The economics of energy storage for infrequent long-lasting lapses of wind or solar generation is untenable, however, since for the capital costs of the storage facility would only be recovered from the high price of electricity during the time spans of rarely occurring weather events.

---

[30] https://holtecinternational.com/communications-and-outreach/smr/

[31] Matsuo, Y. et, al. (2020). "Investigating the economics of the power sector under high penetration of variable renewable energies," *Applied Energy*, 267, 113956.

# Chapter 9
# The Smart Grid's Promise

**Keywords** Grid modernization · Grid structure · Artificial intelligence · Digital systems · Electric vehicles · Rooftop generation · Demand management · Smart thermostats and appliances · Dynamic line rating · Congestion control · Hybrid generation systems · Microgrids · Virtual power plants

"There are places even today where we can't even take one more heat pump without having to rebuild the portion of the system. Or we can't even have one EV charger go in," voiced Palo Alto's Assistant Director of Utilities. The Xcel president told the Minneapolis *Star Tribune* that "four or five families buying EVs noticeably affects the transformer load in a neighborhood, with a family buying an EV adding another half of their house." "Multiple L2 chargers[1] on one distribution transformer can reduce its life from an expected 30–40 years to 3 years," adds an Atlanta energy expert.

The foregoing comments[2] from the front lines of fighting climate change focus on just one segment of the challenges that lie ahead: the overloads on the low-voltage distribution part of the grid triad of generation–transmission–distribution. The increasing penetration of wind and solar energy also requires a massive expansion of high-voltage transmission lines to carry power from remotely located wind and solar farms to the population centers where it is consumed. Dispatchable power generation from nuclear, gas-fired, hydroelectric, and geothermal plants must become more flexible to accommodate large rapid swings in intermittent wind and solar generation; for steady-state operation, it will be replaced with the need for frequent ramping in power as well as operation at small fractions of installed capacity. These challenges must be met even while fossil fueled plants are shut down and the demand for power surges with the growth of artificial intelligence data banks and the electrification of transportation and other segments of the economy to meet the carbon neutral goal of 2025.

---

[1] Level 2 are the chargers most often located in homes.

[2] spectrum.ieee.org/the-ev-transition-explained-2658463709



E. E. Lewis, *Renewables or Nuclear*,
https://doi.org/10.1007/978-3-032-08074-5_9

Two emerging technologies must be central to grid modernization. They fall under the headings of smart grids[3] and artificial intelligence.[4] In the USA, the Energy Independence and Security Act of 2007 defined the smart grid as the "increased use of digital information and controls technology to improve the reliability, security, and efficiency of the electric grid." Artificial intelligence can be applied to many of the smart grid technologies detailed in the pages that follow. It involves creating computer-based systems capable of carrying out tasks that normally require human intelligence and interaction.

The transformation to such a grid would be inevitable even if variable renewable energy (VRE) were not to have a significant place in the future's electricity supply. The need for new generation in the coming decades will accelerate, not only to retire fossil fueled plants but also to electrify other sectors of the economy, such as providing heat and electricity for industrial process and for carbon-free space heating and cooling in venues ranging from homes to large apartment, educational and industrial complexes. Likely the most imminent need for additional generating capacity will come from replacing liquid fueled ground transportation with battery powered vehicles. The replacement will not only cause demand for electricity to grow in magnitude. It will redistribute the demand in time and place, depending on when and where vehicle batteries are recharged.

Earlier grids were one directional, from generation, through transmission, then distribution to consumption. With the advent of rooftop solar generation, however, and the legality of consumers selling their excess power to the distribution utility that serves them, the flow of electricity often reverses on the distribution utility's network of lines. These lines, transformers and other equipment hadn't been designed for reversible flows. Substations connecting high-voltage transmission to distribution must also be reconfigured to deal with flow reversal and other complexities.

Application of state-of-the-art sensors, actuators, two-way communications, and other advanced digital technology to electric grids are defining attributes of what we label smart grids. They are essential in improving efficiency and reliability of grids with or without substantial shares of VRE. Streams of data from the network power plants, substations, transmission and distribution lines, and other facilities may be distilled through the use of artificial intelligence to provide grid operators instantly with the unambiguous knowledge they need to make informed decisions.

Innovations can be implemented across the grid through automated actuators to control power plant output, transmission line loads, and more. Locations of outages from downed power lines or failed transformers will be spotlighted electronically, allowing repair crews to be dispatched immediately. Such data fed into artificially intelligent software will assess the resulting disruption of transmission line network and display—and if mandated execute—rerouting changes in transmission line

---

[3]Kabeyi, M. J. B., & Olanrewaju, O. A. (2023). Smart grid technologies and application in the sustainable energy transition: a review. *International Journal of Sustainable Energy*, *42*(1), 685–758.

[4]https://pmc.ncbi.nlm.nih.gov/articles/PMC11832663/

loadings to compensate for the failed equipment. Likewise, artificial intelligence can quickly compensate for a power plant failure by instantaneously determining which of the grid's other plants should quickly pick up the load, thus avoiding a possible blackout. Equally beneficial may be artificial intelligence utilizing operating, maintenance, and failure data to predict where failures are imminent, thus allowing preventive maintenance to obviate unscheduled shutdowns.

Smart two-way grid communication will be essential as grid penetration of VRE becomes larger, causing whims of the weather to have growing impacts on power generation. Actuation of spinning reserves from dispatchable energy generators to ramp up power quickly may be accomplished electronically. Coordination of power from peaking units with the charging and discharging of battery banks or other energy storage mediums must be automated to supply energy at critical points on the grid to maintain exact balance of supply and demand. Artificial intelligence distilling the vast amounts of data generated by highly automated systems will greatly facilitate grid operates in making timely decisions.

Increasingly, smart grids must allow demand management to offset detrimental effects of weather extremes, equipment failures, and the intermittence of wind and solar generation. Management may take two forms. In peak shaving, the grid operator is given tools to reduce demand in situations where there is danger of it exceeding supply. The cause may be a sudden drop in supply due to equipment failure or a weather induced drop in wind or solar energy. Weather extremes are the foremost cause of demand suddenly exceeding supply, while at the same time causing generators to fail. In load shifting, some part of demand is shifted in time to improve the balance between supply and demand over diurnal cycles, thus utilizing generation and storage resources in the most efficient manner. To implement demand management, however, consumers' homes, businesses, and other establishments must become appropriately equipped, and all stakeholders must be willing to participate in the necessary programs. Here, again, artificial intelligence may be essential for making informed decisions.

***

Before digital electronics allowed electric grids to become more interactive, the rate payers', as electricity consumers were dubbed, interactions with the distribution utility were straightforward, albeit inefficient. Typically, each household had a simple meter, usually located outdoors. Once per month a meter reader would record the number of kilowatt hours (kwh) of electricity on the meter, hopefully without trampling the flowers or allowing the dog to escape. The distribution utility would subtract the previous month's reading and charge for the number of kilowatt hours used, based on the price per kwh—averaged over the month—that the distribution utility charged for electricity. The first customer contact with the emerging smart grid usually comes with smart meters, which allow the provider to charge customers on the basis on cost of electricity at the time it is being consumed, for example, more on a hot summer afternoon when all the generators—from the least to the most expensive—are needed to provide power, and less at night when only the more economical generators are operating.

As utilities advertise this change and encouraged energy conservation, consumers can take common sense measures to reduce their electric bills, dialing back their air conditioners a bit on hot summer afternoons when power is most expensive and moving their use washers, dryers, dishwashers, and other power-consuming appliances to evening or weekend when rates are lower. For most families, space heating and cooling account for a substantially larger share energy use than running appliances. Hence, it is not surprising that the first piece of smart grid apparatus they purchase is the smart thermostat. From a smartphone or computer, such thermostats allow owners to observe the hourly changes in electricity price.

Consumers may program temperature profiles on their smart thermostats to fit their lifestyles and reduce their electric bills. With a cell phone or internet access such adjustment can also be made remotely. For example, they may turn the temperature up before leaving for work in the morning and then turn it back down before leaving for home at the end of the workday, minimizing home air conditioning use in summer and reverse the process to reduce and heating in winter. Increasingly, as smart appliances become available, they may also be programed to function at desired times, or only when the electricity price is less than a chosen threshold.

With increased grid penetration of VRE there will be more days when electricity pricing deviates sharply from typical summer and winter days. Sudden weather fronts may cause wind generation to pick up, causing electricity price to suddenly plummet, but recover more slowly as the front passes. The arrival of a cloud bank can suddenly rob solar farms of much of their power, causing prices to skyrocket. Having wind and solar farms widely dispersed somewhat attenuates such changes, but they still can make gauging when to do laundry, take a long shower or run a dishwasher difficult.

Electric companies welcome the foregoing actions to lower electricity costs, for it is at times of high electricity cost that the grid is most stressed, operating closest to the limits of available generation. Thus, dialing down air conditioning on summer afternoons—or heating down on winter nights—serves as peak shaving, serving to reduce the maximum amount of power that the utility must produce. In most areas, moving appliance use to evening or weekends is a form of load shifting to time spans during which generation capacity is plentiful. On grids that are heavily dependent on solar energy, however, the price may be lowest during the midday hours when solar radiation is most intense, and so the load shift should then be to those hours.

***

Smart grid approaches are essential if the transportation sector of the economy is to switch to electric vehicles (EVs). Understanding where and when drivers will recharge their batteries is essential to providing grid electricity. Possibilities become more expansive when the consumer has battery storage available, whether it be from an EV charger at home, or from a battery unit connected to roof-top solar panels. EV owners may program recharging their batteries when electricity is least expensive, which is also a form of load shifting. For example, the owner may plug into a home charger in the evening and specify how much energy is to be added to the battery by 8:00 a.m. The automated system would distribute the charging through the night to minimize the cost of electricity. Conversely, the distribution utility may pay an incentive to EV owners to draw power from their vehicle battery packs in times of elevated electricity demand. The incentives, however, must be large enough to more than compensate for the shortening of battery life.

In homes, offices and institutional buildings much of the heating has been with natural gas, but electrification is taking hold as a way to fight climate change, with many systems further increasing efficiency by employing heat pumps. For both heating and cooling, a number of approaches may be used to load shift from times when the price of electricity is high to when it is low. A tank hot water heater is a simple example.

Under automated control, when the electricity price is low, smart hot water heaters increase temperature at night to well above a designated value at which it is to be used. As water is used during the day, the temperature in the heater tank will gradually decrease. However, as long as it remains above the designated temperature, a smart valve will mix hot and cool water such that the water leaving the tank will always be at the designated temperature. If on rare occasion so much water is used that the water in the tank falls to the designated temperature, then the heating elements will turn on to maintain that temperature even though the price of electricity may be high.

There are multiple load shifting methods employable for space heating. In apartment complexes with hot water heating, a large storage tank may be heated while electricity prices are low, and the heat distributed through radiators or other means to always maintain temperatures at comfortable levels. Often such thermal storage systems are combined with heat pumps, such as those described in Chap. 6 to further increase energy efficiency. Water is not the only medium for heat storage. For example, in a modern thermal brick furnace, electric heater elements are imbedded in the interior of a ceramic brick with a volume of a few cubic feet. When electricity demand is low relative to supply, the brick is heated to a high temperature, then when heat is needed a fan blows air through channels cut in the brick, carrying heat into the room. A thermostat controls the fan to achieve the desired room temperature. Such systems existed long before smart grids were conceived. However, digital control systems allow coupling them with smart thermostats to vary room temperature with the time of day to coincide with the users' lifestyles.

Load shifting can also be applied to make air conditioning less expensive. The simplest method is applicable to office buildings that are occupied only during standard working hours. There is no additional equipment to be purchased, for the building walls are the thermal storage medium. First a lower and a higher temperature are determined between which occupants will be comfortable. During the night when electricity is least expensive, the air conditioning is run until the lower temperature is reached. During the day, with the air conditioning turned off, the temperature slowly rises from the lower toward the higher value. If the upper temperature is reached, supplemental air conditioning may be needed, but only for a short period of time. Other load shifting systems for air conditioning require a large insulated vessel containing water or other material which is chilled during the night when electric rates are low. During the day, a heat exchanger transfers cool air from the vessel to be distributed throughout the building.

Load shifting may be applied to operations that do not need to be operated continually at a constant rate. Among them are pumping water for irrigation and some operations of water and sewage treatment plants. Many energy-intensive industrial processes that rely on electricity, such as the production of aluminum, may also be shifted from day to night. However, ancillary costs, such as increased nightshift pay, will then be accrued and must be weighed against the load shift benefits. On infrequent occasions when extreme weather conditions or other causes are overloading an electric grid, heavy users in some industries may make arrangements to cut back or shut down. But such arrangements invariably include provision to the industrial consumer to gain recompense through a reduction in the price paid for power.

Load management in the form of peak shaving may be required following the loss of a major power plant or transmission line or the onset of a sudden weather front causing wind or solar power to plummet. On a smart grid, prearranged agreements may allow automatic dialing back or interrupting air conditioning, refrigeration, and/or heating in homes, apartment complexes, schools, or other institutions for a duration limited to assure room temperatures do not vary sufficiently to cause unacceptable discomfort, thawed foodstuffs, stalling of motors powering essential equipment, or other adverse effects. Such actions allow time for other power plants to ramp up and compensate for the lost capacity. With such controls in place, many thousands of electricity end users would be mildly inconvenienced instead of rolling blackouts being imposed. The alternative of rolling blackouts sequentially cutting power off completely to smaller geographical areas has more serious effects of traffic signals going dark, elevators stuck between floors, home medical devices shutting down and more.

We must distinguish between that over which the consumer has control, and that over which the utility does. As the interactions between utilities and consumers grow in complexity, clear transparent understandings must be reached regarding responsibilities, financial remunerations, and limits to protect the interest of all parties. With the necessary agreements in place, additional programs may be implemented to make the grid both more reliable and efficient. Machine learning and artificial intelligence may take the users lifestyle into account to further improve grid operations. Such systems, however, increase privacy issues, and informed consent into what powers are assigned to whom, must be reached between provider and consumer.

***

High-voltage transmission is no less important than generation and distribution in a well-functioning grid, and it faces challenges as electricity demand increases even while the intermittent generation of wind and solar energy cause of transmission lines to operate at lower capacity factors than previously was the case. Thus, on average, the amount of electricity they transport from one place to another is reduced, but at the same time, VRE must be transmitted over longer distances since its generation most often is located farther from the populated areas where electricity use is concentrated. Such circumstances are prone to lead to increased situations where transmission congestion becomes a major problem.

Substantial increases in the capacity of present transmission lines may come from retaining the towers but replacing the lines with advanced materials. The lines capacities could be doubled at a quarter of the cost and over a much shorter time span than building entirely new lines.[5] A second smart grid approach called dynamic line rating (DLR) should also be helpful. New lines are normally rated by the maximum amount of power they can transmit under the worst combination of weather conditions, for example a hot day, with the sun shining and no wind. A DLR system allows more power to be transmitted when weather conditions are more favorable. The system likely employing artificial intelligence would work as follows.

Sensors located along the line relay temperature, solar radiance, wind speed and direction to the grid control center. Automated signals can then be sent to substations to continuously adjust the power that can be safely carried by various lines in a more efficient way. For example, when the temperature is cool, the sky overcast, or the wind is blowing, the transmission line will be allowed to carry more power than its rated value for worst case weather conditions. Other sensors detect short circuits or line breaks, allowing power flows to be transferred immediately to functional lines that have capacity to spare. Banks of batteries and their associated equipment may also be placed at substations to absorb or discharge energy to ameliorate congestion and while power flows are reconfigured.[6]

***

Much of the foregoing deals with the interaction between transmission and distribution of electricity and the end users. Smart grid technology augmented with artificial intelligence should also enable renewable and nuclear generation to deliver electricity through a single connection to the grid while being colocated for carbon-free production of other valuable industrial products such as hydrogen, ammonia, desalinated water.[7] The production of hydrogen has garnered the most attention.[8]

In this case, both wind and/or solar farms and a nuclear reactor feed electricity to a smart junction, part of which goes to the grid, and the remainder to a high-temperature electrolyzer. The reactor also supplies the heat needed for high-temperature electrolysis, which consumes much less electricity than electrolysis carried out at near room temperate. The reactor operates at full power, supplying a constant flow of electricity, while the wind or solar farm output is variable. The smart junction receives continual notice from the grid, indicating the time varying

---

[5] American Institute of Physics, (2024). "Advanced conductors could double power flows on the grid," Physics Today **77,** 6, 21.

[6] Jones, L.E. (ed) (2017) Renewable Energy Integration, Academic Press. Chapter 18.

[7] Bragg-Sitton, Shannon et al. (2014) *Integrated Nuclear-Renewable Energy Systems: Foundational Workshop Report*, Idaho National Laboratory INL/EXT-14-32857.

[8] Ruth, Mark, et al. (2017) *The Economic Potential of Nuclear-Renewable Hybrid Energy Systems Producing Hydrogen*, National Renewable Energy Laboratory NREL/TP-6A50-66764.

amount of power needed to maintain balance, and the junction dispatches it to the grid, while supplying the residual to the electrolyzer.

Note that the VRE farms generate direct current (DC) power which is what the electrolyzer requires, while the reactor generates alternating current (AC) power, which is what the grid requires. Thus, if a VRE farm and electrolyzer are colocated with a reactor, the wind and solar power that goes directly to the electrolyzer circumvents the efficiency losses of the alternator-rectifier sequence for the DC → AC → DC conversion, and only the portion of the reactor power going to the electrolyzer would need to go through the AC → DC rectifier.

The transition between conventional and smart grid interacts with the administrative structure for operating the grid. In particular, it will be different if the grid is under the control of a vertically integrated utility than if it has been split into competing merchant generators, the grid operator, who controls the monopoly network of high-voltage lines, and the distribution utilities that interface with electricity consumers. The primary advantage of the wholesale market arrangements has been the reduction in the cost of electricity. They have also created challenges, including the loss of long-term planning, giving rise to the missing money problem. Another challenge is the loss of communication between power plants and grid control centers. Within vertically integrated utilities there were frequent phone calls between plants to check on how things were going, what needed to be watched in anticipation of problems that might develop and more. But with wholesale market rules such practices typically are forbidden, for fear they could lead to favoritism for one merchant generator over another. Keeping power plants, transmission grid operators, and distribution utilities at arm's length makes the two-way communication between generation, transmission, and distribution more difficult, while such communication is at the heart of successful smart grid operations. Ways must be found to use artificial intelligence such that choice of generators is made on objective criteria that better serve the public than the simplistic marginal cost criterion presently in use.

***

The transformation of today's grids to the smart grids of the future is certain to include two very different developments: microgrids and virtual power plants. While the organization of microgrids[9] varies in detail, generally they may be defined as grids spanning a limited geographical area that can stand alone, operating in "island mode" but in most cases can also be connected to the grid or grids that lie outside their boundaries, thus operating in "grid-connected mode." Within the microgrid's spatial domain, electricity may be generated to fill its needs from both dispatchable and renewable power plants, and substantial battery or other forms of energy storage are likely to be located within the microgrid. Typically, the generating capacity is no more than several megawatts (MW). A low-voltage network distributes power to users within the network. Generally, the microgrid's geographic

---

[9] https://www.energy.gov/sites/prod/files/2016/06/f32/The%20U.S.%20Department%20of%20 Energy%27s%20Microgrid%20Initiative.pdf

area is small enough that high-voltage transmission is needed only to connect the micro grid to the outside grid when it is operating in grid-connected mode.

Microgrids may serve a number of purposes. Economically, they operate in island mode, when it is less expensive to generate electricity internally than to buy it on the surrounding grid, and in grid-connected mode when energy from the surrounding grid is less expensive. More important is the ability to switch to island mode when there is danger of a power outage on the surrounding grid. For this reason, microgrids are often developed to serve large hospital complexes, airports, air traffic control centers, military installations, or other institutions for which it is vital to maintain operations, even though the surrounding grid may be suffering a blackout. The military also finds microgrids invaluable for power in remove areas where the local grid may be unreliable or nonexistent. Future micro nuclear reactors, discussed in Chap. 14, will be transported by land or sea in standard shipping containers to the places where needed to generate electricity and heat for a microgrid.

Stability and balance are as essential in microgrids as they are on the region spanning grids to which they can be connected. As wind and solar energy becomes a substantial share of microgrid generation capacity, a means must be found to replace the synchronous inertia of spinning turbine generators if they no longer exist on the microgrid. Likewise, enough dispatchable energy with fast ramping capabilities and storage must be available on the microgrid to balance supply and demand in the presence of wind and solar intermittency.

Rather than a geographical area such as forms the basis of a microgrid, a virtual power plant (VPP) is a software construct that manages and integrates what are referred to as distributed energy resources (DERs) located at hundreds if not thousands of homes and businesses that may be located over a wide area.[10] Tied together electronically are widely dispersed resources that include many of the devices discussed in the preceding pages: rooftop solar panels, behind-the-meter batteries, electric vehicles and chargers, smart thermostats, smart building temperature controls, hot water heaters, and other appliances.

The VPP is always connected to the grid, and like other power plants sells power into electricity markets. The cloud-based software remotely controls numerous devices to optimize solar energy production, battery storage. Programmable temperature controls of homes and businesses through programming of smart devices and flexible industrial and commercial power demands allow VPPs to provide maximum power to the grid when the need is great. Unlike other power plants, a VPP can also extract energy from the grid when the price is low to recharge its batteries. By doing so, the VPP may serve a similar function to traditional fast-ramping peaking plants to smooth the minute-by-minute and hour-by-hour crests and troughs in residual demand curves, such as those illustrated in Chap. 5. In grids that are heavily dependent on solar energy, however, there is a downside. If overcast skies last for substantially longer than the 4-h discharge times of lithium-ion batteries, the VPP

---

[10] https://www.energy.gov/lpo/virtual-power-plants-projects

batteries will first drain in supplying power to the stressed grid but then may attempt to draw power from the grid, making matters worse.

***

The transition to smart grids will no doubt involve growing use of machine learning and artificial intelligence (AI),[11] for they will have application to many of the challenges that must be faced.[12] The ability of AI to learn from huge amounts of data on weather conditions, distributions of power flows, customer usage patterns and more will give grid operators unprecedented capabilities to keep the grid balanced and stable in the face of diverse sets of circumstances. AI will improve ability to optimize dispatchable energy allocations to complement the weather dependence of renewables. It will provide insights regarding equipment deterioration and point to optimized allocations of preventive maintenance, as well as predict what equipment is likely to fail and during hurricanes, deep freezes, and heat waves. When dealing with contingencies, such as power plant failure or lost transmission lines, AI may offer instant evaluation as how to reconfigure generation, transmission and distribution to stabilize the grid and return it to regular operation. AI uses may include automating demand management decisions related to the peak shaving and load shifting methods described in earlier paragraphs.

Artificial intelligence will be sorely needed to deal with the added complexity of electric grids that stems from adding ever increasing fractions of wind and solar energy generation.[13] Grid operators will need all the AI help available to deal to deal with challenges VRE integration presents. Instead of a smaller number of large power plants supplying dispatchable power to the grid from sites chosen according to where concentrated power consumption is located, many more wind and solar farms will generate from tens to hundreds of MW intermittently from numerous sites. Located according to VRE weather requirements, they are frequently far from where the power is consumed. The configuration of the network of high-voltage lines becomes far more complex and difficult to control. Backup power generators must be strategically interconnected to counter the wide swings in power flow resulting from intermittency, and astute routing changes must be executed on the fly to avoid debilitating line congestions. The operations must maintain grid stability and balance even as the demand for electricity increases, accentuated by the power needs of large AI data banks.

The question arises as to whether machine learning and AI will allow grid operations to be automated completely. The answer must be no, for no software system can be guaranteed to be absolutely bug free. Grid operators must monitor AI output, verify that it is working properly, and take responsibility for implementing actions

---

[11] Pandey, Utkarsh et al. (2023). "Applications of artificial intelligence in power system operation, control and planning: a review," Clean Energy **7**, 1199.

[12] https://www.technologyreview.com/2023/11/22/1083792/ai-power-grid-improvement/

[13] https://www.powermag.com/utilities-grid-operators-grapple-with-adding-renewable-energy/

that it proposes. Smart grids' two-way communications and gathering of customer data also raises ethical questions of protecting privacy, and assuring that information, such as when residents are likely to be away, is not used for illicit purposes. End users must understand what actions distribution utilities are allowed to take on a case-by-case basis without permission, for example, dialing back heat or air conditioning to avoid a rolling blackout, and which are left to the user's discretion, for instance, programing smart appliances to run only when the price of electricity is reasonably low.

# Chapter 10
# On Carbon Containment

**Keywords**  Global carbon cycle · Biomass · Biofuels · Carbon capture · Carbon sequestration · Energy conservation · Deforestation · Ethanol · Direct air capture · Flue gas · Cement and steel decarbonization

The global scope of reducing $CO_2$ concentration in the atmosphere is made clear by Fig. 10.1. It shows that each year hundreds of billions of tons of carbon dioxide (hundreds of $GtCO_2$) are exchanged between the atmosphere, the oceans, and land vegetation. Large annual exchanges existed before anthropogenic emissions became significant, but they were balanced at about 330 $GtCO_2$ between atmosphere and oceans and 440 $GtCO_2$ between atmosphere and land vegetation. In these large $CO_2$ exchanges between the atmosphere and both land and sea, photosynthesis extracts

**Fig. 10.1**  The global carbon cycle. (Source: NASA)



E. E. Lewis, *Renewables or Nuclear*,
https://doi.org/10.1007/978-3-032-08074-5_10

$CO_2$ from the atmosphere to grow terrestrial vegetation and various forms of seaweed. When the plant life dies, the decay process eventually creates $CO_2$, which returns to the atmosphere. Now, however, fossil fuel combustion releases 40 $GtCO_2$ annually. As a result, land and ocean absorption have each increased annual absorption by 10 $GtCO_2$. Therefore, half of the greenhouse gas emissions are reabsorbed through these fortunate feedback effects, and the increase in atmospheric concentration is reduced to 20 $GtCO_2$ instead of the 40 $GtCO_2$ annually that it would otherwise be. Without those fortunate feedback effects, global warming would be far worse!

The emphasis of these pages is on curtailing that 40 $GtCO_2$ by replacing fossil fuels with carbon-free forms of electricity generation and by energy conservation. In contrast, bioenergy is defined as converting vegetation to fuel either for electric power plants, space heating, or automotive transport, while concomitantly the vegetation is replanted. When biofuel is burned, $CO_2$ is released to the atmosphere, but over the life cycle of the vegetation, the same amount of $CO_2$ will be absorbed by photosynthesis. In principle, useful energy has been created in a carbon-neutral process. In reality, the situation is more complicated.

Over the last decade and longer, land-based biomass and biofuel processes, lumped under the heading of bioenergy, have become well established. The original premises upon which bioenergy rests, however, are questionable.[1] Biomass, most often wood, is burned to generate electricity, or the heat is used for space heating other purposes. In principle, an equivalent number of replacement trees are planted to obtain a carbon-neutral process. The argument for carbon neutrality, however, breaks down for both fundamental and practical reasons. Fundamentally, the $CO_2$ will be injected into the atmosphere immediately upon combustion, and the increased $CO_2$ concentration will decrease only slowly from photosynthesis reabsorption, retuning to its initial value only when the new generation of trees has reached maturity. According to plan, the trees will then be cut, their wood combusted, and the next generation of $CO_2$ will enter the atmosphere, repeating the cycle. Hence, during the decades between cuttings, on average roughly half of the $CO_2$ from the previous cutting will be in the atmosphere.

In practice, the situation is much worse. Public relations for bioenergy companies sometimes claim that the wood being burned is primarily the waste from saw and paper mills, furniture factories and similar sources. But growing demand quickly outstripped those supplies. The European Union passed legislation stating that bioenergy is "green." Therefore, it's eligible for subsidies and other advantages, and a huge industry has emerged. Biomass is Europe's largest source of supposed green energy,[2] amounting to more than solar and wind energy combined. Western Europe, however, has few forests left suitable for biomass production. As a result, the continent's last remaining virgin forests are being cut in Romania, Slovenia and

---

[1] https://www.theguardian.com/world/2021/jan/14/carbon-neutrality-is-a-fairy-tale-how-the-race-for-renewables-is-burning-europes-forests?

[2] https://energy.ec.europa.eu/topics/renewable-energy/bioenergy/biomass

other eastern European countries. Moreover, there doesn't seem to be any plan implemented to begin planting replacement trees.[3]

Still, European demand for wood has outstripped supply, resulting in the USA exporting nearly ten million tons of biofuel per year from forests in the southeastern states.[4] A dozen or more mills have cropped up in those states, where whole trees are cut up and compressed into wood pellets. They are then shipped to Europe where they are burned, often in retired coal-fired plants. As a result of the moisture that biomass contains, its combustion emits more $CO_2$ per kilowatt of electricity produced than if coal were burned. Furthermore, the processing of wood into pellets, transport to Europe on ships powered by diesel engines, and grinding the pellets into powder suitable for use in the coal plants releases a great deal of $CO_2$. Such processing and transportation can consume as much as 25% of the energy produced with biomass combustion,[5] and that energy is produced primarily through fossil fuel combustion.

Ethanol is the dominant biofuel in the USA. The federal Renewable Fuel Standard pays farmers to grow corn for ethanol and thereby fight climate change. Consequently, the production and distribution bioethanol has become a large well-entrenched business, with the USA being a world leader in its production. The need for feedstock has kept corn as the most widely planted crop in the USA, with 40% of it going to the production of ethanol.[6] Nearly all gasoline sold contains ethanol, most commonly 10%. But there is a problem. The chemical reactions central to the fermentation processes used to convert corn or other agricultural products to ethanol also produce carbon dioxide.[7] Thus, the sum of $CO_2$ emitted in these fermentation processes plus that produced when the ethanol fuel is combusted is more than was absorbed by photosynthesis. The $CO_2$ added to the atmosphere is further increased by the emissions from farm machinery used to plant, cultivate, harvest and ship the corn and by that generated in the production and application of ammonia-based fertilizers.

Advocates argue that ethanol should be considered "green" because it results in smaller $CO_2$ emissions than using pure gasoline. Even that assumption has been challenged by an extensive study. It concluded that greenhouse emissions are at least equal to those caused by gasoline and probably 24% higher.[8] However, other life cycle studies are contradictory. They indicate that corn ethanol emissions are only about half of those of gasoline. Biodiesel fuel, for which soybeans are the most common feedstock, seems an exception to the shortcomings of burning wood or making ethanol from corn. Life cycle studies have shown that $CO_2$ emissions for

---

[3] https://www.nytimes.com/interactive/2022/09/07/world/europe/eu-logging-wood-pellets.html

[4] https://www.nytimes.com/2021/04/19/climate/wood-pellet-industry-climate.html

[5] https://www.hga-llc.com/wp-content/uploads/2017/06/ENERGY-BALANCE-IN-WOOD-PELLET-MANUFACTURING.pdf

[6] https://www.ers.usda.gov/publications/pub-details/?pubid=105761

[7] https://www.extension.purdue.edu/extmedia/id/id-328.pdf

[8] https://afdc.energy.gov/fuels/ethanol_fuel_basics.html

100% biodiesel are 74% less than those for petroleum diesel.[9] Moreover, biodiesel particulate air pollution is negligible compared to that produced by petroleum-based diesel.

Obtaining useful bioenergy from various forms of seaweed is in early stages of research and development, but it appears to have several attractive properties. Unlike corn or other land-based feedstocks, seaweed requires no irrigation, fertilizers, pesticides, or fresh water. Thus, much less energy is consumed in its farming, and arable land is not withdrawn from food production. Moreover, seaweed grows much faster than terrestrial plants.[10] Giant kelp, which has been most thoroughly investigated, reaches maturity in weeks. Moreover, since it grows year-round, kelp can be harvested every few months.

In coastal waters less than 200 meters deep, upwelling causes deep water to bring nutrients close to the surface, accelerating photosynthesis in the presence of sunlight and $CO_2$ from the atmosphere. Further from shore in the deep ocean, the surface waters are nutrient poor to beyond the depths that sunlight penetrates. Accordingly, more advanced farming methods are being explored to extend seaweed production over wider stretches of ocean.[11] Concomitantly, studies are exploring several chemical processes to determine whether seaweed may be used as a feedstock to produce green methane, hydrogen and possibly aircraft fuel.[12]

As present, the energy from fossil fuels that is expended in producing either biomass for electric generation or biofuels for transportation is resulting in net increases of $CO_2$ to the atmosphere. The processes may be carbon neutral in principle, but they are far from it in practice. Improved technology, and possibly the use of seaweed feedstocks, may reduce the net amount of $CO_2$ released, but cannot eliminate it completely. The prospects for bioenergy of all varieties improve greatly, however, if they can be combined with the carbon capture and sequestration that we discuss in the pages that follow. When combined with $CO_2$ capture and sequestration, bioenergy applications should result net extractions of $CO_2$ from the atmosphere.

***

Carbon capture and sequestration (CCS) provides a valuable tool in reducing the 40 $GtCO_2$ released to the atmosphere if it is used to remove $CO_2$ from the flue gases of fossil and biofueled power plants. Indeed, CCS has been used for decades to remove $CO_2$ and $H_2S$ impurities from natural gas, and more recently research has been directed toward using the process to scrub $CO_2$ from the flue gas emissions of coal and methane fueled power plants.

Carbon capture is widely accomplished with amine scrubbing, which is a chemical process.[13] Flue gas that contains $CO_2$ is mixed with an amine, a class of organic

---

[9] https://www.caranddriver.com/research/a31883731/biodiesel-vs-diesel/

[10] https://www.scientificamerican.com/article/could-our-energy-come-from-giant-seaweed-farms-in-the-ocean/

[11] https://dornsife-wrigley.usc.edu/news/is-seaweed-the-future-of-flying/

[12] ibid.

[13] Herzog, H. M. (2018). *Carbon Capture*, MIT Press.

compounds to which $CO_2$ adheres. The mixing takes place in an absorber tower with the hot flue gas rising from the bottom and the amine descending from the top. The tower contains a packing that enhances contact between flue gas and the amine solution. The flue gas—no longer containing $CO_2$—is exhausted to the atmosphere. The $CO_2$, chemically bound to the amine, then enters the top of a second tower, called a stripper. High-temperature steam enters the bottom of the tower and strips the $CO_2$ from the amine. The amine is recycled back to the absorber tower. Upon leaving the stripper column, the steam is condensed, leaving a stream of nearly pure $CO_2$ to be compressed to liquid form as the first step in sequestration.

Typical $CO_2$ concentrations in flue gas are about 12%. Amine scrubbing generally can remove more than 90% of the $CO_2$ from the gas. In this process, there are two major operating costs. The first is for the energy consumed in providing the steam to the stripper. The second is for the electricity consumed at the end of the process to compress the $CO_2$ to liquid form suitable for transport or sequestration. For carbon capture from most power plants, the needed steam can be extracted from the steam turbine. Adding carbon capture to a natural gas combined cycle plant, however, reduces the net electricity it generates by about 15%, and at present nearly doubles its capital cost.[14]

The concern with global warming has caused a resurgence in research in carbon capture, and progress is being made. Improvements in removing $CO_2$ from flue gas range from replacing amines with better solvents to substituting other configurations for the towers. Rotating beds eliminate the need for steam, and some employ cryogenics. The objectives range from increasing efficiency to well over 90% capture to driving down capital costs and energy penalties. More radical departures relate to precombustion processes. Coal gasification was once considered a promising prospect, but results to date have been disappointing. Another technology called Oxy-Combustion Capture deserves a look.[15]

Oxy combustion takes advantage of the fact that capture is most effective when there is a high percentage of $CO_2$ in the flue gas. Cryogenic separation is used to obtain pure oxygen from air, which normally contains 78% nitrogen. Accordingly, if coal or gas is combusted with pure oxygen, the concentration of $CO_2$ will be much higher in the flue gas since no nitrogen will be present. Therefore, carbon capture will be more efficient. The tradeoff will be in whether the increased efficiency of carbon capture will more than compensate for the added expense of the cryogenic oxygen separation and other power plant modifications.

Bringing down carbon capture's cost is a necessity if it is to gain widespread use in eliminating greenhouse gases from fossil and biofuel combustion. The greatest need, however, may be to capture the carbon from the exhaust of gas-fired power plants that ramp up and down rapidly to complement the intermittent power generation of wind and solar farms. The rapidly varying exhaust streams from the gas

---

[14] Ibid.

[15] Ibid.

turbines in such situations are likely to be more difficult to decarbonize than for those emanating from the same turbine operating at steady state.

Carbon capture may have uses that go well beyond reducing greenhouse gas emissions from fossil fueled electric generators and from bioenergy combustion and fermentation processes. The largest releases of $CO_2$ from the industrial sector come from the production of steel and cement. Both are presently produced principally by high-temperature coal combustion. But even if coal could be replaced by nuclear, hydrogen, or some other carbon-free heat source, cement production requires a chemical reaction that produces $CO_2$. Thus, carbon capture is the only means available for preventing the $CO_2$ generated in the production of cement from entering the atmosphere.

***

Capturing carbon from flue gas created by power plants, from industrial processes or directly from the air—as discussed in the following section—is only a first step. The captured $CO_2$ must be cooled to less than 30 °C (87 °F) and compressed at a pressure of 74 atmospheres to reduce the gas to liquid form. Such compression accounts the largest fraction of the energy consumed in carbon capture and sequestration. Finding a use for the $CO_2$ and transporting it to a place from which it will not leak to the atmosphere for thousands of years are the remaining tasks. For distances of a few hundred miles, tanker trucks can transport tens of tons of liquid $CO_2$. For the millions of tons per year, that must be captured to have an appreciable impact on climate change; however, pipelines would be needed for cross country transport. Large ships similar to those used for transporting liquefied natural gas may be suitable for oceanic transport.

More than 4000 miles of pipeline presently operate in the USA to deliver $CO_2$ to oil and gas wells, where it is inserted to enhance recovery.[16] They exist because the $CO_2$ sources are far from the wells. Presently, over half of that used in wells comes from naturally occurring underground $CO_2$ caverns. The remainder comes from carbon capture. Using carbon captured from flue gases as a feedstock for enhanced oil recovery prevents an equal amount of $CO_2$ from being extracted from natural reservoirs. That makes the process carbon neutral. In fact, nearly all of the $CO_2$ injected in the wells remains there, and that which is pumped out with the oil and natural gas is recycled back into the wells. Thus, it is a carbon sink. Nevertheless, because reducing the need for fossil fuels is a goal of most projects for fighting climate change, using captured carbon to enhance oil recovery to many seems antithetical.

According to a Department of Energy study, there is sufficient underground storage capacity in the USA to sequester hundreds of years of domestic $CO_2$ emissions.[17] However, the reservoirs must remain leak proof for 10,000 years or more, and that is a challenge. They must be porous with good permeability so that the liquid $CO_2$ will readily fill the pours. Storage must be well over a half mile below

---

[16] https://www.energy.gov/policy/articles/review-co2-pipeline-infrastructure-us

[17] https://netl.doe.gov/carbon-management/carbon-storage/faqs/carbon-storage-faqs

the surface, for only then will the ambient pressure be more than the pressure needed for the $CO_2$ to remain a liquid. Even then, since liquid $CO_2$ is buoyant in water, impenetrable capstone must separate the reservoir from the surface. Deep saline reservoirs filled with salty water seem most suitable. Depleted oil and gas reservoirs may also be acceptable, but only if the wells have not compromised the integrity of the capstone.

Carbon capture for the most part will take place at power plants, steel mills or cement kilns. These are unlikely to be located conveniently close to suitable reservoirs. As a result, the expense of creating a suitable pipeline infrastructure may be excessive. The challenges parallel those of creating the natural gas pipelines. However, the two differ in that the methane has only to be stored for months before it is used, and it has value at the diverse points of delivery, while the carbon dioxide has none; it is simply to be irretrievably stored. If salt domes or other suitable storage sites are developed beneath the seas, then pipelines might converge on major seaports where the $CO_2$ could be transported by ship to storage. Norway is currently developing such a site, along with agreements for shipping the $CO_2$ there by sea.[18] Costs no doubt will determine whether such an arrangement is economically viable.

Proposals have also been made for sequestering $CO_2$ at great depths in the oceans.[19] Because pressure increases with depth, at about 2 miles beneath the surface, liquid $CO_2$ becomes denser than water and therefore should sink and form a lake on the ocean floor. However, too little is known about conditions at those depths, including the long-term stability of $CO_2$ deposits, and their effects on ocean chemistry and marine biology. Thus far, availability of domestic underground storage and the concern for unintended consequences have ruled out the detailed consideration of deep ocean sequestration.

***

The path forward is to eliminate fossil fuel emissions from the generation of electricity and then, by electrification, to reduce greenhouse gas emissions from the transportation, industrial, and other segments of the economy. Carbon capture likely will play a significant role in decarbonizing electricity generation through adding equipment to capture carbon from the flue gases of coal- and gas-fired generators. Beyond that, however, there are segments of the economy for which it is next to impossible to eliminate nearly all greenhouse gas emissions. Agriculture is a prime example, but aviation and ocean shipping, and several aspects of commerce also fall within this category.[20]

Diminishing returns from mitigating greenhouse gas emissions eventually takes hold as the remaining sources become increasingly difficult to mitigate. Thus, to achieve carbon neutrality, $CO_2$ removal from the atmosphere will need to be widely employed. A number of paths have been put forward,[21] some proven, others only in

---

[18] https://ieefa.org/articles/norways-carbon-capture-and-storage-projects

[19] https://www.sciencedirect.com/topics/engineering/deep-ocean-storage

[20] https://www.ox.ac.uk/news/2021-08-05-final-25-how-tackle-hard-reach-emissions

[21] National Academies, (2015). *Climate Intervention, Carbon Dioxide Removal and Reliable Sequestration*.

**Fig. 10.2**  The Orca, Iceland direct air carbon capture plant. (Source: Orca)

the early stages of research. Of the possible means, one of the most effective would seem to be bioenergy combined with carbon capture and sequester to make the process strongly carbon negative. However, since land requirements could limit the expansion of bioenergy, other approaches are being considered. At present, the most advanced of these is direct air capture (DAC):

September 2021 was a celebrated occasion, headlined in newspapers, and carried on TV. It was the opening of the world's largest plant for extracting $CO_2$ directly from the atmosphere.[22] Located on a lava plateau in Iceland, the Orca facility, shown in Fig. 10.2, is designed to remove 4000 tons of $CO_2$ from the atmosphere per year. Heralded by commentators as a major step forward in the fight against climate change and for obtaining carbon neutrality by 2050, DAC has an intuitive simplicity for cleaning the atmosphere directly, not present in the slow slog to mitigate the greenhouse gas emissions from fossil fueled power plants. Orca's innovative design fascinated engineers, and the fact that it was totally green made it environmentalists' dream fulfilled.

The plant's design was innovative in several respects.[23] Like other plants, it employs large fans to blow air through where the $CO_2$ adheres to a solvent. But there the similarity ends. Orca does not use towers for extraction. Instead, the plant consists of a wall of modules, each of which can be manufactured off site and then added to the complex as needed. In this respect, it parallels the move in the nuclear industry from constructing large reactors on site, to building small modular reactors manufactured primarily in factories. The modular structure also allows Orca to employ a new solid adsorber, with the modules going through on–off cycles: turned

---

[22] https://climeworks.com/news/climeworks-launches-orca

[23] https://techcrunch.com/2021/12/03/co2-capture-iceland-climeworks-orca/

on for the absorber to trap the $CO_2$ from the air, and off while the $CO_2$ is scrubbed from the absorber. This is quite different from the standard continuous carbon capture process in which liquid adsorber traps the $CO_2$ in a settling tower, and the mixture then flows to where the adsorber is scrubbed from the $CO_2$, before the adsorber is returned to the settling tower.

Orca is uniquely "green." It utilizes electricity from a nearby geothermal plant to power its fans, heaters and other equipment, and to sequester the $CO_2$. Just as important, the local geologic conditions allow the captured $CO_2$ to be mixed with water and pumped deep into basalt bedrock. There in about 2 years it should mineralize into a solid carbonate, a rock similar to limestone.[24] The DAC process as implemented at Orca may at first seem to be an ideal tool for combating climate change. But is it? To answer the question, we must first place DAC and climate change within the broader perspective of the global carbon cycle and examine alternate methods for reducing the presence of $CO_2$ in the atmosphere.[25]

The idea of direct air capture or DAC as implemented at the Orca plant is appealing in its simplicity, but its widespread adoption at this juncture would not be as cost effective as applying carbon capture to flue gases of fossil fueled power plants or industrial processes. The reasons are straight forward. Most importantly, the atmospheric concentration of $CO_2$ is only 0.04%, while the flue gas coming from the stacks of coal- or gas-fired plants averages about 12% and can be significantly higher.[26] Thus, for the same amount of $CO_2$ captured, a DAC plant must process 12/0.04 = 300 times as much atmospheric air as the flue gas processed in a fossil fuel plant. As a result, for the same amount of $CO_2$ captured, much more energy is expended with DAC than with flue gas capture; estimates are that DAC cost may be more than ten times that of flue gas capture per ton of $CO_2$.[27]

For the Orca plant in Iceland and other plants being constructed to capture carbon directly from the air, the transport and sequester challenges can be resolved by locating DAC at sites above a suitable reservoir. Colocating small modular or micronuclear reactors to provide decarbonized power would negate the need for high-voltage transmission lines stretching from distant renewable energy or nuclear power plants to DAC sites.

Bioenergy with carbon capture and DAC are not alone as methods for extracting $CO_2$ from the atmosphere. Several other approaches are under active investigation.[28]

---

[24] Ibid.

[25] https://environmentamerica.org/missouri/center/resources/carbon-dioxide-removal-the-right-thing-at-the-wrong-time/

[26] Herzog, op. cit.

[27] https://www.technologyreview.com/2018/04/24/66958/the-daunting-math-of-climate-change-means-well-need-carbon-capture/

[28] https://nap.nationalacademies.org/catalog/18805/climate-intervention-carbon-dioxide-removal-and-reliable-sequestration

Energies Futures Initiative, (2022). *CO2-Secure, A National Program to Deploy Carbon Removal at Gigaton Scale.*

Among them are reforestation, changes in agricultural practice, ocean fertilization, and enhanced weathering. Their cost effectiveness, scalability, and environmental consequences, however, are not as well understood as those of carbon capture and sequestration. Examination of their long-term prospects is outside the scope of this volume.

# Chapter 11
# Energy from Einstein's Equation

**Keywords** Nuclear power · Nuclear fission · Reactor development history · Naval reactors · Electric power reactors · Political and economic interactions · Today's reactors · Capital cost reduction

Albert Einstein's equation $E = mc^2$ relating energy to mass and the speed of light is arguably the most celebrated formula in the modern world. Among its more widespread applications is the analysis of the energy created and transformed in nuclear power reactors. That energy is huge, measured in millions of electron volts (MeV). At those levels, the mass changes accompanying nuclear reactions are measurable, unlike the much smaller changes that take place in chemical reactions, which are measured in electron volts (eV). The defining reaction related to nuclear reactors is fission, which is pictured in Fig. 11.1.

In the diagram, a neutron collides with the isotope uranium-235, creating uranium-236, which is unstable and fissions, creating about 200 MeV of energy, two or

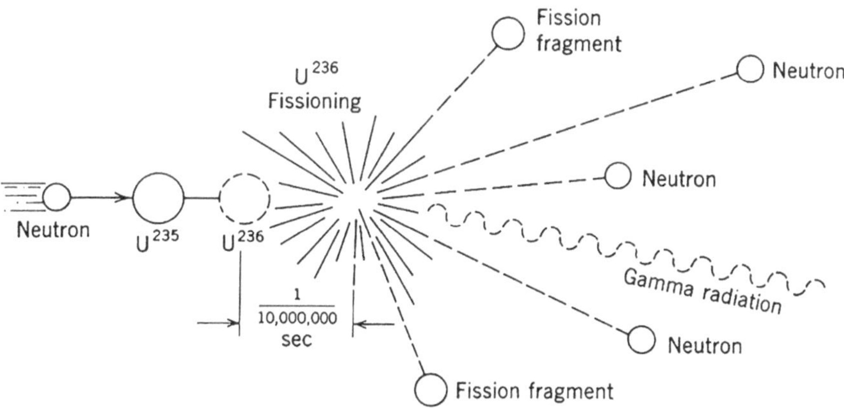


**Fig. 11.1** Nuclear fission of uranium-235. (Source: Self drawn)




E. E. Lewis, *Renewables or Nuclear*,
https://doi.org/10.1007/978-3-032-08074-5_11

three neutrons with average energies of 2 MeV, and two highly radioactive fission fragments—one usually substantially heavier than the other. Adding the masses of the neutrons and fission fragments—gamma rays have no mass—yields a mass that is less than that of uranium-236 by an amount equivalent to 200 MeV of energy, as given by Einstein's equation. In contrast, the energy released in fossil fuel combustion is in the eV range. In burning coal, for example, carbon is oxidized, according to $C+O_2 \rightarrow CO_2$, yielding carbon dioxide, and about 4.0 eV of energy. Therefore, fission releases 50 million times as much energy as coal combustion.

Leading to fission was a sequence of discoveries that began near the turn of the twentieth century. In 1896, Henri Becquerel found that uranium salts spontaneously emitted radiation. In 1898, Marie and Pierre Curie expanded the study to included radiation emitted by thorium and polonium as well as uranium. Marie Curie termed the phenomena "radioactivity." Shortly thereafter Ernest Rutherford observed that there were three distinct emissions, which he designated as alpha, beta, and gamma radiation. The 1903 Nobel Prize was awarded to Becquerel and the Curies for their work on radioactivity. Of the three types of radiation, the alpha radiation, which was found subsequently to consist of helium nuclei, played a central role in the progression of events that led to the discovery of fission in the 1930s.

In 1932, James Chadwick discovered the neutron; they were emitted when he bombarded a sheet of beryllium with alpha particles. Then in 1934, Irene Joliot-Curie and her husband Frederic produced artificial radioactivity by bombarding aluminum foils with alpha particles. They called it artificial because it was not spontaneous but rather was emitted from a material which was not naturally radioactive. The same year Enrico Fermi published his seminal paper describing how irradiating heavy elements with neutrons caused new isotopes to form, most of them radioactive. The series of discoveries climaxed in December 1938 when Otto Hahn and Lise Meitner announced that by bombarding uranium with neutrons, they caused it to fission, and most importantly that neutrons—more than one—resulted from the reaction. Chadwick, the Curies, Fermi and Hahn all received Nobel Prizes for their work, but sexism and antisemitism prevented Meitner from receiving the prize to which she had every right.

Following the 1939 Nobel Prize Ceremony in Sweden, Enrico Fermi didn't return to his professorship at the University of Rome. He, his Jewish wife, and their children emigrated to the United States to escape persecution in fascist Italy. He joined the faculty of Columbia University and continued his research. By then he realized that his 1934 neutron experiments had produced fission, but much to his regret, he had not recognized it. His quest turned to bringing about the first neutron chain reaction.

Slow neutrons—with energies less than 1 eV—are what Hahn and Meitner used to cause fission, but the neutrons produced by fission were fast, with an average energy of about 2 MeV. Thus, Fermi must pair uranium with another material that would slow the neutrons produced by fission down to where they could cause fission, releasing the next generation of fast neutrons. He needed a low atomic mass material, for neutrons colliding with light atomic weight nuclei would lose much more energy as they scattered in various directions than if they scattered from the nuclei of heavier atoms. (Think of the difference between a billiard ball hitting another billiard ball and it hitting a bowling ball.) Hydrogen, and hence water, seemed the perfect material to slow down neutrons.

The result, however, was disappointing. Through carefully conceived experiments, Fermi showed that water was effective in slowing neutrons down, but then too many of them were then absorbed by the water; not enough would remain to produce fission and thus the next generation of neutrons. Fermi and his students examined other substances and settled on carbon in the form of graphite. With an atomic mass of 12, it was less effective than water in slowing neutrons down, but his experimentation showed that carbon absorbed very few slow neutrons. Hence, he chose graphite to be the moderator, moderation being the name he gave given to the process of slowing down neutrons; slow neutrons became known as thermal neutrons.

Fermi's understanding of nuclear physics, statistical mechanics and more allowed him to design experiments to answer crucial questions: what should be the ratio of uranium to graphite? Should the uranium be interspersed in lumps or be powdered? How big must an assembly of uranium and graphite—or the pile as he dubbed it— be? Based on theoretical considerations, Fermi chose lumps of uranium placed in holes drilled in a matrix of the graphite blocks to make up the pile. In other slots bored in the graphite, he placed rods made of cadmium, which he had determined absorbed neutrons. His experimental program consisted of carefully pulling out the cadmium rods and making measurements on a series of piles, each larger than the last, for the larger the pile, the smaller would be the fraction of neutrons escaping its surface. With each experiment, Geiger counter readings of the neutron population would grow to a higher level, but then stabilize. From the readings, he could calculate the pile's multiplication, defined as the ratio of neutrons in one generation to those in the preceding generation. The letter $k$ was, and is, used to designate multiplication. With a large enough pile, $k$ should equal one, indicating a sustained neutron chain reaction, and the pile would be said to be critical. To this day, reactors for which the multiplication, $k$, is less than one, equal to one, or greater than one are said to be subcritical, critical, or supercritical.

Before Fermi's experimental piles were large enough to reach criticality, the onset of World War II interrupted his progress. In Europe and the USA, the realization came that nuclear fission could lead to the creation of an atomic bomb of horrendous destructiveness, and the first nation to create it would have the military superiority to end the war on its terms. The race was on. Fermi's lab was moved to Chicago and made part of the Metallurgical Laboratory, a code name to keep what became the Manhattan Project a secret. Fermi's experimentation continued at a feverish pitch. Most of the more serious delays stemmed from problems obtaining large quantities of graphite free of impurities that would absorb neutrons.

The laboratory was a doubles squash court under Stag Field, an abandoned football stadium on the campus of the University of Chicago. Chicago Pile 1, or CP-1, consisted of 288 lumps of uranium evenly distributed in holes drilled in the $4 \times 4 \times 12$ inch graphite bricks. Day and night shifts carefully stacked layer upon layer of

**Fig. 11.2** The first nuclear reactor. (Source: Chicago Historical Society)

the bricks to form a nearly spherical pile, 25 feet high and 20 feet wide, supported by a wooden structure.[1]

On December 2, 1942, 48 scientists under the direction of Fermi witnessed the historic Experiment from the squash court balcony. The neutron absorbing rods were withdrawn from the pile one by one, the last inch by inch, with Fermi making slide rule calculations after each Geiger counter reading. Then the Geiger counter clicks turned to a buzz, and on a chart a pen recorded the neutron population in the pile rose at a faster and faster pace. Finally, Fermi raised his hand and announced that "Pile has gone critical." A value of $k = 1.0006$ had been reached. Reducing the value of $k$ to 1.0000, he then let the chain reaction continue for 2.5 min at a steady power of one-half watt before he ordered it shut down. Pictured in Fig. 11.2, the pile had produced the first self-sustaining controlled neutron chain reaction. Thus, the nuclear reactor was born.

Achieving a controlled neutron chain reaction was a historic achievement, but at that time it seemed of little value to the objective of the Manhattan Project. Since slow neutron fission required a moderator to achieve criticality, the size of an explosive device would be far too large to use as a bomb. Independent of Fermi's efforts, experiments showed that fast neutrons also caused uranium to fission. However, no bomb could be built with the naturally occurring Uranium that Fermi had employed, for it contained only 0.7% of fissionable uranium-235, with the remaining being uranium-238. Consequently, a mass of natural uranium, without a moderator, could not become critical, no matter how big it was; the uranium-238 would absorb too many of the neutrons.

---

[1] https://www.osti.gov/opennet/manhattan-project-history/Events/1942-1944_pu/cp-1_critical.htm

Manhattan Project experiments, however, showed that fast neutron could also cause uranium-235 to fission, and theory indicated that if the isotope could be separated, then a uranium-235 bomb based on fast neutron fission could be made with only about 100 of pounds of the uranium-235 isotope. However, since isotopes cannot be separated by chemical means, enriching uranium to contain a high percentage of uranium-235 was going to be a prodigious time-consuming task. To pursue the uranium bomb option, massive gaseous diffusion plants were being constructed to do the enrichment.

Meantime another option had opened. Glen Seaborg in Berkeley, California, produced a new element by bombarding uranium-238 with neutrons. They were absorbed, producing uranium-239. It, however, was unstable and underwent radioactive decay over several hours, producing the new element called plutonium-239. Instead of it having an atomic charge of 92, like uranium, it had a charge of 94. Further study showed that it, like uranium-235, was fissionable with either fast or slow neutrons. To distinguish the differing properties of the two uranium isotopes, uranium-235 was designated as *fissile*, since it is fissionable, and uranium-238 as *fertile* since if it absorbs a neutron, becomes unstable, and undergoes radioactive decay to become plutonium-239, which is also fissile. Plutonium was a different element and hence could be separated from uranium chemically. Plutonium's critical mass was only about 22 pounds, much less than that for uranium-235, and it could be produced in a graphite-moderated reactor. After all, Fermi's achieving the first controlled neutron chain reaction contributed to the war effort.

***

Following World War II, several countries developed graphite-moderated reactors, usually cooled by gas, to generate electricity. They were fueled with natural uranium and concentrated in nations without adequate facilities for enriching uranium. Heavy water, made up of deuterium (one proton and one neutron) instead of hydrogen (one proton), could also be used as a moderator and coolant for natural uranium reactors since it absorbed few neutrons, but it was very difficult to produce. In the USA, ordinary—also called light—water became the coolant of choice, for uranium fuel enriched to only a few percent of uranium-235 could compensate for light water's absorption of neutrons. Moreover, water cooled reactors fueled with slightly enriched uranium were more compact than reactors fueled with natural uranium.

In the USA, the naval officer and engineer Hyman Rickover was an early proponent of nuclear propulsion for ships, recognizing that fission energy could power ships for months or years without refueling, eliminating the need for oil tankers following the navy's fleets for frequent refueling. To minimize the reactor size and weight, naval design combined highly enriched uranium fuel, with pressurized water serving as booth coolant and moderator. Under Rickover's leadership, research began in 1948, and in 1955, the Nautilus, the first nuclear powered submarine, set to sea. First submarines and then aircraft carriers were built using nuclear reactors for propulsion. Nearly 300 reactors on the Navy's ships have accumulated 7500 reactor years of operation, with the more recent ships designed to sail their entire lives without refueling.[2]

Impatient with private industry's reluctance to make the investments needed to bring about a civilian nuclear power industry, President Eisenhower called on Admiral Rickover's naval propulsion program to adapt a design originally intended to power an aircraft carrier to produce electric power instead.[3] Naval expertise was invaluable in building the first nuclear reactor in the USA to generate sizable amounts of electricity: Begun in 1954, the Shippingport reactor in Pennsylvania started delivering power to the electric grid in 1958. Cooled by pressurized water, it was built by the Westinghouse Corporation under the auspices of Admiral Rickover's naval propulsion program. It was operated by the Duquesne Light and Power Company. Subsequently, private industry took the initiative. The Commonwealth Edison Company contracted with General Electric to build the Dresden 1 reactor in Illinois. Commencing operation in 1960, it was a boiling water reactor, based on government funded research carried out at Argonne National Laboratory in the 1950s.[4]

The depth of experience gained in the design and operation of water-cooled reactors caused water to become the coolant of choice for electric power reactors. Minimum cost rather than minimum size and weight was the key criterion for commercial power reactors. Hence, uranium fuel of low enrichment of 2–4% was used rather than the more expensive highly enriched fuel needed for naval reactors. Pressurized water and boiling water reactors (PWRs and BWRs) soon dominated the US market. In the 1970s and 1980s, more than 100 of these light-water-cooled reactors (LWRs) came online to supply 20% of electricity demand of the USA. As low enrichment uranium became more available internationally, other counties switched from graphite-moderated reactors, fueled with natural uranium, to employing LWRs fueled by slightly enriched uranium.

The widespread deployment of LWRs across the world did not come without challenges. Stress corrosion cracking, steam generator tube blockages, and a variety of fuel failure mechanism caused frequent unscheduled shutdowns, some of them quite lengthy to understand the faults and correct them. It took years before the capacity factors of the LWRs operating today were reached.[5] During the 1970s, when many reactors came online, they were about 50%. Following the Three Mile Island 2 accident in 1979 improved operational practices in the following decades, led to capacity factors reaching 70% or higher by the turn of the century. A steep rise then brought them above 90%, where they remain today.

***

---

[2] https://www.navy.mil/Press-Office/News-Stories/Article/3476093/naval-reactors-celebrates-75-years/

[3] Hewlett, R. G. and J. M Holl,(1989). *Atoms for Peace and War 1953-1962, Eisenhower and the Atomic Energy Commission,* University of California Press.

[4] https://www.osti.gov/servlets/purl/4115425/

[5] https://www.nei.org/resources/statistics/us-nuclear-generating-statistics

Figure 11.3 displays schematic diagrams of a pressurized water and a boiling water reactor, worldwide the predominant reactor types for producing electricity; pressurized water reactors captured 85% of the market. In a pressurized water reactor, heat is produced by nuclear fission in the thousands of fuel rods that constitute the reactor core. Each is a zircaloy tube containing cylindrical pellets of the ceramic uranium dioxide ($UO_2$). Heat from nuclear fission is transferred from the fuel elements to the water coolant.

In the pressurized water reactor on the left of Fig. 11.3, coolant circulates out of the pressure vessel and to the steam generator where the heat is extracted; the coolant is then pumped back to the reactor. The coolant is kept at too high a pressure for it to boil: 220 atmospheres pressure at more than 320 °C (600 °F). The steam generator piping transfers the heat to water at a low enough pressure for it to boil. The steam is then piped to the turbine which in turn drives the generator. After passing through the turbine, the steam is condensed to water and recycled back through the steam generator. The neutron absorbing control rods regulate the power at which the reactor operates.

The boiling water on the right of Fig. 11.3 differs from the pressurized water reactor primarily in that the water boils in the reactor core, and there is no steam generator. Instead, the steam exits the reactor pressure vessel and powers the turbine directly. The pressure and coolant temperature are lower, 75 atmospheres and 285 °C (545 °F), respectively, than in the pressurized water reactor. The lower pressure allows a thinner pressure vessel and piping to be used in a boiling water reactor.

Building on the extensive experience gained by the navy in developing water cooled propulsion reactors and the increased availability of enriched uranium, the USA excelled in expanding its number of LWRs in 1970s and 1980s to become the world's leading generator of electricity from nuclear fission. Then, however, the industry went into decline, and for more than 30 years no reactors were built, and only now are there signs of recovery. The causes for decline were threefold. First, lingering nuclear fear was evoked by the accidents at Three Mile Island and

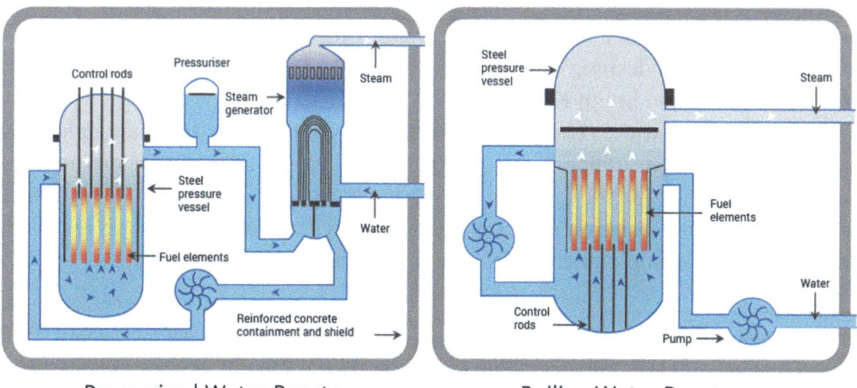


Pressurized Water Reactor                    Boiling Water Reactor

**Fig. 11.3** Pressurized water & boiling water nuclear reactor layouts. (Source: World Nuclear Association)

Chernobyl. Second, the 1990s' economic deregulation of electricity markets in large parts of the country made it exceedingly difficult to raise adequate capital to build large power reactors with cost recovery periods of a half century or longer. And third, the precipitous price drop caused by the rise of fracking technology made natural gas-fueled power plants less expensive to build and operate than nuclear reactors.

***

For three decades, top- and mid-level management retired from the nuclear industry and from the Nuclear Regulatory Commission. Likewise, the pool of tradespeople and other workers tuned to the challenges of nuclear construction vanished, as did much of the domestic equipment supply chain. Many major components, such as pressure vessels, were no longer manufactured domestically. The Southern Company confronted this situation in 2009 when it announced plans to add two Westinghouse AP1000 reactors as units 3 and 4 to its Vogtle power plant site in Georgia. The designs were much improved over earlier reactors, particularly the safety systems. The safety systems rely on natural convection, eliminating the need for many pumps and valves that might fail under accident conditions.[6]

Construction of the two 1.2 gigawatt reactors began in 2009. The first was to connect to the grid in 2016 and the second in 2018. Since then, the original cost estimate of 14 billion dollars for the two units has more than doubled, and the schedule has slipped by 7 years. Vogtle-3 only came online in 2023, and Vogtle-4 was completed in 2024. Consequently, these reactors are frequently used as an argument that reactors cannot be considered part of the solutions to the climate crisis: They are too expensive, and they take too long to build. But is that true?

It should be no surprise that the two Vogtle reactors have encountered large cost overruns and schedule delays. The causes are many and varied, some of them stem from the lack of experience with reactor construction in the USA for more than three decades, while others date back to the earlier period when cost was less of a consideration.[7] Comparison of US practices to those abroad during the Vogtle reactors' construction shows paths toward reducing construction costs and eliminating time delays.[8] The difficulties do not stem from the reactor technology itself, for reactors of identical design, Westinghouse AP1000s, have been built on budget and on time in China and South Korea.

In western countries, cost overruns and time delays have stemmed primarily from the indirect costs of engineering, procurement, and construction. Commencing with construction before detailed design is complete and resolving regulatory issues contribute to cost escalation and construction delays while problems are resolved, and workarounds implemented. Outdated project management practices lead to

---

[6] https://www-pub.iaea.org/MTCD/publications/PDF/P1500_CD_Web/htm/pdf/topic3/3S05_P

[7] Eash-Gates, et. al. (2020). "Sources of Cost Overrun in Nuclear Power Plant Construction Call for a New Approach to Engineering Design". Joule 4, 2348.

[8] https://www.oecd-nea.org/jcms/pl_30653/unlocking-reductions-in-the-construction-costs-of-nuclear

problems with subcontractors' late deliveries and faulty components. Moreover, nearly 50% of the cost escalation and schedule slip came not from the reactors or turbine-generators but from antiquated on-site construction practices of the reactor containment and other nonnuclear excavations and buildings. Late materials deliveries caused work stoppages resulting in cranes and other expensive equipment to sitting idle and in labor productivity plummeting.

The foregoing deficiencies are being analyzed and methods implemented to reduce costs and speed construction.[9] Such should be the case, for example, with advanced methods for steel-concrete composite construction. The lessons learned are from experience with large gigawatt sized reactors, but some methods for lowering costs and shortening construction times should also be applicable to the small modular and advanced reactors discussed in Chap. 14. Moreover, most of the reactors built in the 1970s and 1980s are still operating with capacity factors of 90% or more. Their operating licenses are being extended from 40 to 60 years, and consideration of 80-year licenses has begun.

Thus, high costs notwithstanding, in his 2021 book, Bill Gates unabashedly makes a one-sentence case for nuclear power.[10] "It's the only carbon-free energy source that can reliably deliver power day and night, through every season, almost anywhere on earth, that has been proven to work on a large scale." Why is it then that the public is reluctant to endorse nuclear energy for a major role in the fight against climate change. Aside from high cost, it presents fears of nuclear accidents, waste, and weapons proliferation, fears that we shall address in the following chapters.

---

[9] National Academies (2023). *Laying the Foundation for New and Advanced Nuclear Reactors in the United States.*

[10] Gates, B. (2021 ) *How to Avoid a Climate Disaster*, Vintage Books, P84.

# Chapter 12
# Radiation and Radiophobia

**Keywords** Inherent radiation fear · Hiroshima and Nagasaki · Radiation releases · Psychological impacts · Three Mile Island · Chernobyl · Fukushima · Cancer predictions · Actual accident consequences

The Cuban Missile Crisis erupting in October 1962 gave my wife and me the scare of our lives. Only a few months into our marriage, the world seemed on the brink of nuclear holocaust. Russian ships were making their way toward Cuba with missiles that could hit Chicago, New York, and other cities in the eastern USA only a few minutes after launch. President Kennedy had blockaded Cuba, thus giving Nikita Khrushchev an ultimatum. If war broke out, Chicago, a hundred and fifty miles away, was likely the nearest target to where we were. Certainly, we thought, the Soviet Union did not have enough missiles to target Champaign, Illinois. Distance would protect us from the blast, heat, and radiation from a bomb's detonation. Our fear was the fallout from the ensuing radioactive cloud if it should pass overhead, made worse if rain deposited radioactive debris on our surroundings, contaminating the environment long after the cloud had passed.

From my studies, I knew that two types of radiation would be a threat: beta particles and gamma rays. Beta radiation consists of very high energy electrons. It is nonpenetrating. Hence, just keeping the windows closed would offer excellent protection. Danger would only come from ingestion or inhalation, allowing radioactive debris to concentrate in susceptible organs, radioactive iodine in the thyroid and strontium and cesium in bone marrow. Gamma rays are the same as X-rays, but with much higher energy and therefore more penetrating. In the absence of shielding thicker than the lead aprons in dentists' offices, gamma rays would penetrate the body and spread their damage uniformly. Hence, exposure to gamma radiation is often termed a "whole body dose." In a basement, the earth would provide substantial gamma shielding, but our grad school apartment didn't have one. Short of that, we planned to shelter in our apartment, out of sight from the windows, behind a closed door, thus allowing the wooden door, brick walls and the floor, ceiling, and roof of the apartment above us to offer some shielding. We jammed emergency



E. E. Lewis, *Renewables or Nuclear*,
https://doi.org/10.1007/978-3-032-08074-5_12

groceries—anything that wouldn't spoil unrefrigerated—under the bed, bought flashlights and a battery powered radio, and cleaned and filled buckets with water in case our supply was cut off. For the next week and more, we were glued to our radio as the crisis grew, and we tried to stay near to where we could quickly shelter in place before the radioactive cloud reached Champaign.

Only when Khrushchev blinked, turning his ships away from Cuba, did the tension break. Khrushchev and Kennedy then worked out a diplomatic solution in which Khrushchev would not install missiles on Cuban soil, and Kennedy agreed not to invade the island. During those 13 days, virtually everyone experienced intensified fear of radiation. But it was a fear that had been being imprinted since the bombings of Hiroshima and Nagasaki, 17 years earlier, and to some extent it continues to be etched in our subconscious to this day.

On August 6, 1945, the Bombing of Hiroshima created a mushroom cloud that rose into the stratosphere to a height of more than 60,000 feet, while the city was engulfed in a thick black smoke. The bombings of Hiroshima and Nagasaki riveted attention on the horrendous destructive power from splitting the atom, heretofore unknown to the public. The Hiroshima and Nagasaki bombing resulted in 66,000 and 39,000 deaths, respectively, and in both cities, there were nearly as many injuries.[1] The vast majority of deaths from the two catastrophes were caused by blast and heat, rather than by radiation. Nonetheless, what understandably scared everyone around the world most was the radiation unleashed by the explosions. Unlike the blast and heat, which quickly came to an end, the cloud carried radioactive debris high into the atmosphere, and there it would linger and spread far beyond the stricken cities. Gruesome pictures published later showed the effects of acute radiation sickness had on survivors close to the blasts, many of whom died within weeks. Though unsubstantiated, the fear of delayed effects of cancer and birth defects soon began to grow.

Following World War II, international tensions dominated the news. The Soviet Union exploded its first atomic bomb in 1949 at a test site in Kazakhstan, and fear grew that the Soviet Union would rain atomic bombs on the USA. Civil Defense programs encouraged the building of bomb shelters, and the Duck and Cover program had school children practice crawling under their desks if they saw the blinding bright light that would signal an atomic explosion. Others protested such drills, objecting to the idea that nuclear war was survivable.

In 1951, the USA commenced atmospheric testing at Yucca flat in Nevada. Fallout from the tests was readily detectable as it spread across the county causing widespread fear of radiation induced cancer, birth defects, or other maladies. The Atomic Energy Commission, the government agency overseeing the tests, lost credibility by asserting that the exposures were too small to have health effects. A political movement to ban atmospheric testing formed and grew rapidly. It gained added impetus in 1952 from the much larger than predicted energy release of the first hydrogen bomb. It was detonated above Eniwetok Atoll in the Marshal Islands,

---

[1] https://www.atomicarchive.com/resources/documents/med/med_chp11.html

three thousand miles west of Hawaii. Two years later, a second test on Bikini Atoll turned out to be the largest bomb ever exploded, three times larger than predicted; it was a thousand times more powerful than the bomb dropped on Hiroshima.

Meanwhile, as if constant news coverage of nuclear testing wasn't enough, a growing list of doomsday novels, movies, and even comic books fed on the public's fear of radiation.[2] Dr. Strangelove, a deranged scientist, and Godzilla, a radiation empowered monster, were pop culture favorites. Fictional accounts of a nuclear World War III implied that civilization would end. Often, these emphasized not the blast and heat that was the predominant cause of death at Hiroshima and Nagasaki, but of lingering health threats thought to be caused by radioactive fallout.

Typical of the emphasis on radiation contamination was the 1957 novel by Nevil Shute, *On the Beach*.[3] It described an imagined World War III in 1963. The blast and heat of atomic war irradiated all life in the northern hemisphere. But it was not the sudden death of much of the world's population that most horrified the readers. Rather, the book focused on the spooky radioactive fallout that then crept into the southern hemisphere. The plot centered on its dire effects. Australians, spared the immediate nuclear destruction, waited for inevitable extermination as trade winds caused deadly radioactive fallout eventually to cover their country, causing extinction of their society. The book sold more than a million copies and horrified movie goers in its 1959 film adaptation. Such was the fear of radiation at the time of the Cuban Missile Crisis.

Following resolution of the Cuban Missile Crisis, Cold War tensions subsided somewhat. The crisis spurred the creation of the Hot Line, a link between Moscow and Washington DC providing a way for the two Cold War opponents to negotiate crises' settlements by communicating directly. Soon thereafter, they negotiated a treaty to ban all nuclear testing from the atmosphere, from space, and from the oceans. It was signed in Moscow on August 5, 1963. By then, however, the fear of radiation had been deeply implanted in the public psyche, and there it would remain, stoked by extensive media coverage of any mishap, large or small, related to radiation or radioactive materials.

***

On March 28, 1979, a mechanical failure and faulty instrument readings combined with operator errors caused the nuclear chain reaction at the Three Mile Island 2 nuclear reactor to shut down. The combination of faults, however, disabled the cooling required to remove the heat produced by the decay of radioactive fission products accumulated in the reactor fuel. As a result, much of the reactor core eventually melted. Even so, the radioactivity was confined within the reactor's containment structure, which was built to prevent radioactivity from leaking into the atmosphere under such circumstances. No member of the public was exposed to

---

[2]Weart, Spencer R., (2012).*The Rise of Nuclear Fear*, Harvard University Press.
[3]Shute, Nevil, (1958).*On the Beach*, New American Library.

measurable amounts of radiation because of the accident, and even those working at the plant were adequately protected.[4]

Nevertheless, the citizenry was understandably horrified. Congress demanded investigations, and President Carter visited the plant to calm the public. Subsequently, federal regulations were tightened, and utilities reorganized to increase safety. Orders for nuclear reactors were canceled, and work on plants under construction was suspended. Operating plants were equipped with added layers of protective equipment, and operators underwent greatly beefed-up training. Eventually, nuclear plant construction resumed, and the share of electricity coming from the reactors grew to 20% of US demand.[5] In some areas of the country, it was much higher. In the Commonwealth Edison service area, which included Chicago and its suburbs, nearly three quarters of the electricity came from nuclear fission.

Seven year later, on April 26, 1986, as the public was beginning to be less fearful of operating reactors, a far worse accident took place in Ukraine at the Chernobyl Nuclear Plant.[6] The reactor never would have been licensed to operate in the West. Its outdated Soviet design was known to be unstable under some operating conditions, and it had no containment structure to protect the public if the worst should happen. Furthermore, the crew was negligent, performing unauthorized experiments that it should have known were dangerous. Making matters worse, the operators had turned off the rudimentary safety systems that the reactor did have in order to perform their illicit experiments.

As a result, an explosive power surge was set off that destroyed the reactor core and blew the roof off the building housing it. The blast ignited a graphite fire that consumed a large part of the reactor core. The fire's heat sent a highly radioactive plume into the atmosphere, spreading contamination over large areas of the western Soviet Union and Europe. Contamination of the area surrounding the plant was most severe. Acute radiation exposure took the lives of 29 plant workers and sickened many more as they attempted to bring the fire under control. In the following months, over 160,000 residents were evacuated from the communities nearest the reactor, and eventually, even more were relocated. The Soviet government was complicit in the disaster. It tried first to hide and then to downplay the accident's severity. As the radioactive cloud spread over much of Europe, it was detected first in Sweden and then elsewhere. Panic ensued, and dire predictions were made.

In the wake of the Chernobyl calamity, a media frenzy spread dubious claims that tens or even hundreds of thousands of cancer deaths would result from the radiation released. Fear of birth defects ran rampant. Organizations, such as the Greenpeace and the Union of Concerned Scientists, used radiation risk models to predict tens of thousands of cancer deaths across Europe, and even further.[7]

---

[4] https://www.energy.gov/ne/articles/5-facts-know-about-three-mile-island

[5] https://www.eia.gov/energyexplained/nuclear/us-nuclear-industry.php

[6] OECD (2002). *Chernobyl: Assessment of Radiological and Health Impacts*. Report 3508.

[7] https://blog.ucsusa.org/lisbeth-gronlund/how-many-cancers-did-chernobyl-really-cause-updated/

**Fig. 12.1**  Fukushima tsunami destruction. (Source: iStock)

Radiation fear set off a world-wide scramble for iodine tablets to provide protection from thyroid cancer. The entire crop of reindeer meat in Sweden was destroyed, and sheep in England were prevented from being sold because they had grazed on contaminated fields.

Twenty-five years after the Chernobyl disaster, on May 11, 2011, a record-breaking earthquake of 9.0 on the Richter scale burst forth 80 miles off the eastern coast of Northern Japan. A tsunami of 40 feet ensued, overtopping and destroying the sea wall built to protect Fukushima prefecture. As illustrated in Fig. 12.1, the wave brought unimaginable death and destruction, washing away buildings, drowning inhabitants, leaving nearly 20,000 dead or missing![8]

The earthquake was detected, and the six nuclear reactors at the Fukushima-Daiichi power plant were shut down before the tsunami hit. The reactors were of western design, equipped with safety systems and containment structures. Nevertheless, flooding of emergency equipment inhibited the removal of the heat produced by the radioactive decay of fission products in the reactor cores, and three of the cores melted, and there followed hydrogen explosions that caused their containments to be breached. As a result, radioactive materials were released to the atmosphere. Authorities mandated evacuation zone to the 150,000 people living within 12 miles of the reactors. Two weeks later, the government enlarged the evacuation zone and recommended that an additional 350,000 residents relocate further from the plant.[9]

---

[8] OECD (2021). *Benchmark Study of the Accident at the Fukushima Nuclear Power Plant* Report 7525.

[9] https://world-nuclear.org/focus/fukushima-daiichi-accident/fukushima-daiichi-accident-

Radiologically, the Fukushima accident wasn't nearly as severe as that at Chernobyl; it spewed about a tenth of the radioactive material into the atmosphere. Nonetheless, as with the Chernobyl debacle, predictions of thousands of cancer deaths were spread by the media, and panic ensued even for many who lived an ocean away.[10] Imminent death wasn't the issue, but dread of how many would perish in the coming years from the cloud of radioactive material spreading over the surrounding countryside, contaminating soil and sea, and carrying detectable amounts of radioactive isotopes to other countries and other continents. Round the world nonstop TV coverage stretched into weeks, with experts, including the author, frequently appearing to give commentary on their understanding of what was transpiring, while trying to damp down unfounded fears. In reaction to public sentiment, governments not only in Japan, but also in Germany and elsewhere, chose to shut down reactors and discontinue further nuclear development.

Both the Chernobyl and Fukushima accidents did, indeed, release large quantities of the most volatile fission products, namely, radioiodine, strontium, and cesium. Several other volatile elements were released but have such short half-lives that they do little harm. Strontium and cesium with half-lives of about 30 years are the most prominent long-term forms of radiological contamination, some of it settling to the ground near to the reactor, the remainder being dispersed in the atmosphere over much more expansive areas. The uranium and even heavier transuranic elements, mainly plutonium, are involatile and were released in only trace amounts.[11]

More than three decades have elapsed since the Chernobyl explosion and more than a decade since the Fukushima disaster. With one exception, none of the predictions of large numbers of excess cancers, birth defects, and other maladies have materialized, not even among those living close to the stricken reactors.[12] An excess of 6000 thyroid cancers followed the Chernobyl disaster. Even these were preventable if children had been told not to drink the milk or eat food contaminated by radioactive fallout. Only 15 deaths resulted, however, for thankfully thyroid cancer can be treated successfully.[13] Public health measures prevented any thyroid cancer following the Fukushima reactor meltdowns, and there have been no deaths attributed to radiation in the disaster's aftermath.[14] Two reactor operators drowned, and one died from the collapse of a crane.[15]

Why did the cancer epidemics predicted to stem from the Chernobyl and Fukushima accidents fail to materialize? The cancerous effects from radiation exposures at Hiroshima and Nagasaki certainly had. Media coverage is partially to

---

[10] https://world-nuclear.org/information-library/safety-and-security/safety-of-plants/fukushima-daiichi-accident

[11] https://www.cnsc-ccsn.gc.ca/eng/resources/health/health-effects-chernobyl-accident/

[12] https://www.unscear.org/docs/reports/2008/11-80076_Report_2008_Annex_D.pdf

[13] https://world-nuclear.org/information-library/safety-and-security/safety-of-plants/chernobyl-accident.aspx

[14] https://world-nuclear.org/information-library/safety-and-security/safety-of-plants/appendices/fukushima-radiation-exposure.aspx

[15] Ibid.

blame. It played on the radiophobia that had been instilled in the public since those World War II bombings, exaggerating the perception of health hazards of accidental radiation releases. Worse yet, those exaggerations were given credence by scientifically flawed prognostications made by otherwise reputable organizations. In good part, the failure to make meaningful assessments of radiation risk originated from the dubious use of Linear no threshold or *LNT* hypothesis, which we shall examine in the following chapter.

# Chapter 13
# Regulating Risk

**Keywords** Radiation risks · Regulation history · Threshold regulation · Linear no threshold regulation · Comparative risks · Fallout · Atomic bomb survivors · Chernobyl · Fukushima · Hypothetical accidents · ALARA · Over regulation

Radiological safety has been a concern since long before the bombs were dropped on Hiroshima and Nagasaki in 1945. It dates back to shortly after W.C. Roentgen announced his discovery of X-rays in 1895. He named them X-rays because their nature was unknown. Soon after, hospitals employed the first crude X-ray machines to diagnose broken bones. Medical use moved quickly beyond broken bones to locating bullets and swallowed objects, and before the turn of the twentieth century, chest X-rays were used to diagnose tuberculosis. As X-ray use became widespread, skin burns, eye irritations, and other problems indicated that health hazards stemmed from unbridled exposures. A nongovernmental body, the International Commission on Radiation Protection (ICRP), organized and, in 1931, adopted the concept of "tolerance dose" as an upper limit for safe exposure. Subsequently, the "roentgen (R)" was defined as the amount of radiation producing a specific amount of ionization in tissue, and in 1934, the ICRP established a tolerance dose of 0.2 R per day of exposure. In 1936, the ICRP reduced this limit to 0.1 R per day.[1] Thus, the ICRP firmly established the concept of a threshold below which usage was safe.

Illustrated in Fig. 13.1, the threshold concept became near universal in dealing with dosages of both radiation and chemical toxins. It made sense in that many chemicals known to be dangerous in high doses are harmless or even beneficial at low does. Aspirin and other pharmaceuticals taken for pain relief are beneficial in small amounts, while massive doses lead to acute illness or death. Below a threshold chlorine in drinking water does no harm while providing the benefit of killing bacteria, viruses and other disease-causing microorganisms. If the concentration were to be increased beyond the threshold, however, adverse health effects would appear

---

[1] https://www.ncbi.nlm.nih.gov/books/NBK232703/


E. E. Lewis, *Renewables or Nuclear*,
https://doi.org/10.1007/978-3-032-08074-5_13

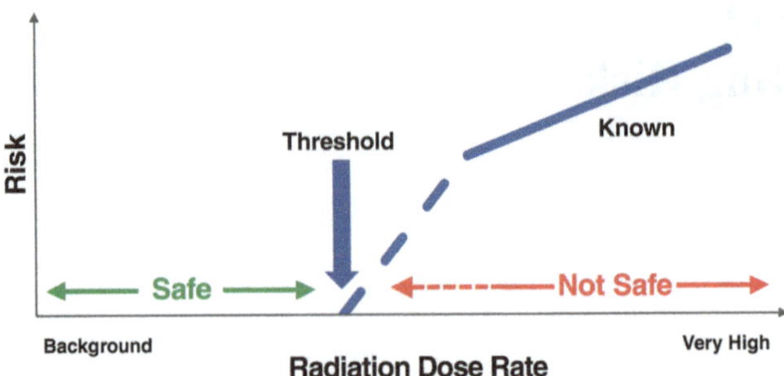


**Fig. 13.1** The threshold dose response model. (Source: Health Physics Society)

and increase with exposure. In fact, highly concentrated chorine gas was used as a chemical weapon in World War I.

In the years before World War II, studies subjecting fruit flies to high X-ray doses led the geneticist Hermann J. Muller to hypothesize that the number of radiation-caused mutations was proportional to the dose, no matter how small the dose. Hence, he initiated the linear no threshold (*LNT*) dose–response model for the biological damage caused by radiation.[2] Even though his experimental data on which *LNT* was based was later shown to be invalid, Muller was awarded a Nobel Prize in 1946 for the *LNT* model. He then began a campaign to have *LNT* adopted as the universal model for regulating radiation risk.

The *LNT* model is diametrically opposed to the threshold criterion that it replaced. The model states that no matter how small the radiation dose, there will be cancer risk proportional to its intensity; accordingly, cancer risk from radiation only vanishes if there is no radiation exposure. The straight line of Fig. 13.2 represents the *LNT* model. The horizontal axis is usually measured in Sieverts (Sv) or millisieverts (mSv). The harm on the vertical axis is most often a probability, such as of contracting cancer or a congenital birth defect. The straight line is drawn from data obtained from very large doses (indicated by the solid line) and extrapolated to very low doses (indicated by the dashed line). Muller's work extrapolated doses to millions of times smaller than those from which the data was obtained. The probabilities of cancer are so small at very low doses that it is impossible to obtain large enough statistical samples to prove or disprove the results.

To many scientists at that time, the *LNT* model didn't—and still doesn't—make sense. It's not consistent with what is readily observable. Background radiation exists everywhere. It is variable but universal, coming from trace amounts of uranium, thorium and other radionuclides in rock formations, Radon in the air, and cosmic rays. In the USA, average background exposure is 3 mSv/year, but rock formations and cosmic rays cause it to be three times as much on the Colorado

---

[2] Calabrese E. J. The road to linearity: why linearity at low doses became the basis for carcinogen risk assessment. *Arch Toxicol.* 2009;83(3):203–.

https://www.sciencedirect.com/science/article/pii/S0009279718311177?via%3Dihub#fn16

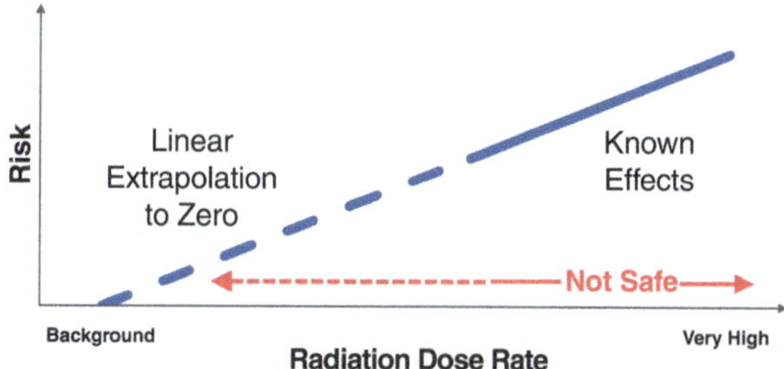


**Fig. 13.2** The linear no threshold dose response model. (Source: Health Physics Society)

Plateau than in the Gulf States. Yet, cancer rates on the Plateau are lower than near the Gulf of Mexico.[3] Other variables such as diet, genetics, or other environmental factors likely play more dominant roles.

Other evidence of *LNT*'s misrepresentation of radiation risk abounds. Even at the locations where background radiation is near its highest, such as on the black sand beaches of Brazil, where radiation is over ten times the 3 mSv/year. US average, studies have since found no correlation to increased cancer rates.[4] Animal studies, such as those using many thousands of mice at Oak Ridge National Laboratory, failed to show increased cancer risk for radiation doses less than 100 mSv.[5] The gold standard of human studies has lasted nearly 75 years. It has carefully tracked the more than 94,000 survivors of the atomic bombings along with a matched control group of 27,000 participants who were not present on the days of the bombings. Doses were estimated based on position and shielding for each person when the bombings occurred. Survivors who received high doses exhibited increased cancer mortality and shortened lifespans. Those who were exposed to smaller doses, however, showed decreased cancer mortality and longer lifespans than the control group, thus providing evidence of some hormesis (i.e., favorable health effects) from low doses of radiation.[6]

In addition to the epidemiological, animal, and human studies, current understanding of radiation biology also exposes the weakness of the simplistic logic implied by the *LNT* model.[7] The model is based on the single hit theory. It assumes that a single DNA ionization caused by an X or gamma ray can cause cancer, and that there is no repair. If that were true, then the ionizations would accumulate, increasing the probability of cancer in linear proportion to the number of ionizations. Moreover, it then would follow that as long as the total radiation dose is the same, it doesn't matter whether the radiation damage is accumulated in a short burst

---

[3] https://pubmed.ncbi.nlm.nih.gov/9753369/

[4] https://pdf.blucher.com.br/physicsproceedings/v-ecfa/020.pdf

[5] Calabrese, op sit.

[6] Sutou, S. (2018) *Genes and Environ* **40**, 26.

[7] https://link.springer.com/article/10.1007/s00204-013-1104-7

or spread over a long period of time—a conclusion that is verifiably false. It is akin to stating that the effects of taking a sleeping pill once per day for a year would have the same effect as taking 365 of them at one time. The *LNT* hypothesis also forces the conclusion that 10 people subjected to 1 Sv and 10,000 people subject to 1 mSv (i.e., 1/1000 of a Sv) would likely result in the same probable number of cancers.

The results of many carefully executed studies led to near consensus among radiation specialists: A threshold should be established at about 100 mSv for a radiation dose received over short period of time. For chronic doses, the corresponding threshold should be about 200 mSv/year.[8] Below these thresholds, radiation exposure has no adverse effects. In fact, there is considerable data as well as cellular and evolutional biological arguments to be made in favor of radiation hormesis, that is of beneficial effects of radiation expose at less than the threshold values.[9] But the threshold pardine was not established. Instead *LNT* was adopted and is still used by governmental agencies around the world for regulating radiation. The reasons have more to do with political influence than with science. The story begins not long after Muller was awarded his Nobel Prize.

***

In 1953, President Eisenhower addressed his historical Atoms for Peace Speech to the United Nations. In it, he proposed applying nuclear energy to essential needs, including abundant electrical energy. It inspired numerous young people, including the author, to make careers of the peaceful uses of nuclear energy. But the fossil fuel industry viewed it from a quite different perspective. During the Eisenhower presidency—1953–1961—political battles raged over whether the government or private industry should develop the nuclear power industry.[10]

The electric utility industry, and those who fueled it with coal and oil, were upset that the naval reactors program under Admiral Rickover's direction built the first electric power reactor at Shippingport, Pennsylvania. They fought the federal program of building additional demonstration reactors around the country, fearing that new federal agencies similar to the Tennessee Valley Authority would displace private industry in the generation of electricity. Eisenhower favored private industry developing nuclear energy. Industry, however, wanted to lead the development, but only if it received such large subsidies that it need not make the large capital investments required to build the reactors. Instead, the utilities opted for the status quo of fossil fueled electricity generation.

During those same years, the radioactive fallout from atmospheric testing of nuclear weapons was of widespread concern, and an unbiased evaluation of the biological effects of the radioactive fallout was sorely needed.[11] Closely aligned to

---

[8] https://hps.org/publicinformation/ate/faqs/regdoselimits.html

[9] https://www.ncbi.nlm.nih.gov/pmc/articles/PMC2477717/

[10] Hewlett, R. G. and J. M Holl,(1989). *Atoms for Peace and War 1953–1962, Eisenhower and the Atomic Energy Commission,* University of California Press.

[11] Ibid.

the oil industry, the Rockefeller Foundation waged a successful campaign on behalf of the *LNT* model. At the behest of the foundation, President Eisenhower agreed to have a Committee on the Biological Effects of Ionizing Radiation (BEIR) appointed by the US National Academy of Science, a prestigious nongovernmental organization, to study the issue and make recommendations.

The effort leading to *LNT* came from the Genetics Panel of BEIR, and suspicions of the panel's bias were many.[12] The panel's chair was the long-term director of scientific research at the Rockefeller Foundation, and the research support of the BEIR members came through him. Muller, a vociferous advocate for his *LNT* theory, was on the panel. The deliberations were highly questionable, with three of nine panel member's risk estimates deleted for no justifiable reason, and for years the chairman refused to publish the panel's deliberations. Nevertheless, in 1956, the panel's recommendation led to *LNT* becoming the guiding principle for federal regulation of carcinogenic risk of radiation and chemicals, and as such it has remained.[13]

The *LNT* predictions are understandably scary in that no matter how small the radiation dose, a risk of cancer is still present. This played into the hands of the fossil fuel industry in its efforts to stymie the growth of nuclear power in the USA. It did so not directly, but through surrogates. In 1973, Robert O. Anderson through his Atlantic Richfield Foundation targeted nuclear energy by founding Friends of the Earth with a $200,000 grant.[14] Fossil fuel interests also gave millions of dollars to other antinuclear organizations such as Sierra Club and the Environmental Defense Fund.[15]

More than half a century has transpired since the *LNT* model was adopted by regulatory agencies in the USA and by many other nations as well. Over the last decade, however, support has grown to replace the simplistic "single hit model" with threshold regulation based on a more nuanced understanding of radiation damage at a cellular level and the body's response to it. The argument for change is expressed succinctly by the position statement PS010-4 of the Health Physics Society, a scientific organization of professionals working in the field of radiation protection[16]: *Recent low-dose research indicates that biological response mechanisms such as DNA repair, bystander effects, and adaptive response modulate radiation-induced changes at the molecular level. Cellular transformation leading to carcinogenesis by mutation of genetic material appears to be a complicated, multistep process that is not reflected in the LNT model.* The Health Physics Society then elaborates that of most concern is that the extrapolation of the *LNT* model to low

---

[12] https://www.ncbi.nlm.nih.gov/pmc/articles/PMC2477717/
   https://www.cato.org/regulation/spring-2019/troubled-history-cancer-risk-assessment

[13] Ibid.

[14] Engdahl, F. W. (1992). A Century of War, Anglo-American Oil Politics and the New World Order, edition.engdahl.

[15] https://environmentalprogress.org/the-war-on-nuclear

[16] Position Statement of the Health Physics Society PS010-4: Radiation Risk in Perspective, *Health Physics* 118(1):79–80, January 2020.

doses and the use of collective dose over large populations as a metric for risk. Other highly respected scientific organization have abandoned the *LNT* model for assessing radiation risk as well. They include The French Academy of Science, The United Nations Scientific Committee on the Effects of Atomic Radiation (UNSCEAR), and International Commission on Radiological Protection (ICRP).

***

Fear of radiation made the radiation consequences of the Chernobyl and Fukushima disasters seem much worse than they were. Exaggerations of radiation risk propagated through rumor, sensationalized press coverage, and inflated predictions of tens of thousands of cancer deaths resulting from radiation exposure. At the heart of those predictions was the use of the *LNT* hypothesis to estimate collective dose, the very fallacy that the Health Physics Society and other professional societies now warn against. Moreover, even if those studies were to be believed, the collective radiation doses were spread over such huge populations that the increase in cancer occurrence would have been an undetectably tiny fraction of the cancers already occurring in the absence of the added radiation.

Over the weeks following the Chernobyl disaster, the population of 116,000 evacuated from the area around the plant received and average dose 30 mSv, and an average dose was 9 mSv in the three contaminated countries: Belarus, Russia, and Ukraine, which is roughly equivalent to the dose from one CAT scan.[17] Therefore, except for thyroid irradiations, no doses to the public approached the 100 mSv generally agreed to be the threshold for radiation risk. Several international studies have concluded that elevated rates of depression, suicide and other psychological issues rather than radiation exposure were the dominant public health effects. Indicative of the mental health crisis following the Chernobyl accident, the International Atomic Energy Agency estimated 100,000–200,000 wanted pregnancies were aborted in Western Europe because physicians mistakenly advised patients that the radiation from Chernobyl posed a significant health risk to unborn children.[18]

The argument is sometimes made, that even if *LNT* exaggerates low-level radiation risk, its use in regulation is justified: Its "conservatism" provides an extra margin of safety, and that is a good thing. But is it? Not necessarily, for measures taken to reduce one risk disproportionately may cause other larger risks. The Fukushima disaster is a case in point. Analysis has shown that no one—not even those living in the evacuated areas close to the reactors—would have suffered a radiation dose approaching 30 mSv/ year,[19] much less than half of the threshold dose considered safe by health professionals. A World Health Organization report stated that the average dose to those living in

---

[17] Ibid.

[18] https://jnm.snmjournals.org/content/jnumed/28/6/933.full.pdf

[19] https://world-nuclear.org/information-library/safety-and-security/safety-of-plants/appendices/ fukushima-radiation-exposure.aspx

the Fukushima prefecture was about 20 mSv, which is only about double the background radiation, and well below the 100 mSv threshold.[20]

Excessive radiation fear, however, resulted in the evacuation of more than 150,000 residents from the vicinity of the Fukushima reactors. Hospitals and nursing facilities were emptied, medications disrupted, transportation improvised, and housing for many ill or aged adults was secured only in overcrowded schools, gyms and other public facilities. The physical and mental stress of evacuation caused an estimated 2313 deaths.[21] Food, water, and inadequate care resulting from the evacuations were causes, as well as fatigue, chronic disease and suicides. Many deaths could have been avoided if a program for simply sheltering in place instead of evacuations had been implemented.

Because they retain the *LNT* model, federal regulatory bodies such as the Nuclear Regulatory Commission and the Environmental Protection Agency do not set thresholds such as the 100 mSv thought to be reasonable by an increasing number of professional organizations across the world. Instead, *LNT*-based policy in the USA led to the "As Low as Reasonably Achievable" (ALARA) principal, established in 1954 by the National Council on Radiation Protection and Measurements.[22] ALARA is based on the idea that any amount of radiation exposure, big or small, can increase negative health effects, such as cancer, for an individual. The principle often causes undue fear of X-rays and other medical imaging[23] as well as of reactors and other peaceful uses of the radiation.[24]

ALARA implementation often results in exposures being reduced to a tiny fraction of the 100 mSv guideline, often at great cost in building and operating nuclear plants. Expensive construction delays are often caused by frivolous lawsuits because design and operational decisions are difficult to defend in court against the subjective ALARA criteria. The small extra margin of safety that ALARA criteria may provide must also be weighed in economic terms of costs, delays and even cancellations of nuclear plants. A case in point is the NuScale reactor, the first of the new small modular reactors, and one that differs least in design from existing power plants. The NuScale reactor recently went through the process to obtain a certificate of design from the Nuclear Regulatory Commission (NRC). It cost more than 500 million dollars and two million labor hours, and the 12,000-page application took nearly 4 years to process.[25] An additional site-specific NRC license will be required before the plant can be build and operated.

Figure 13.3 makes obvious the negative safety impacts of slowing the transition away from fossil fuels by delaying nuclear power generation for small and often

---

[20] https://www.who.int/news-room/questions-and-answers/item/health-consequences-of-fukushima-nuclear-accident

[21] https://world-nuclear.org/information-library/safety-and-security/safety-of-plants/fukushima-daiichi-accident.aspx

[22] https://www.cdc.gov/nceh/radiation/alara.html

[23] https://www.ncbi.nlm.nih.gov/pmc/articles/pmid/32425724/

[24] M. Lips, et al. (2021). J. Radiol. Prot. 41, S306.

[25] https://world-nuclear-news.org/Articles/NuScale-SMR-receives-U.S.-design-certification

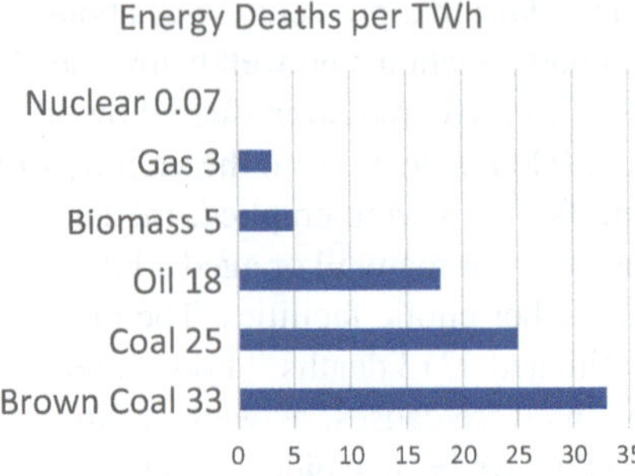


**Fig. 13.3** Energy deaths per TWh. (Source: ourworldindata)

questionable reductions in radiation exposure. These statistics show that compared to burning fossil fuels, the health risks of nuclear power are miniscule. Burning coal results in tens of thousands of deaths each year from asthma and other lung-related diseases, and oil is not far behind. Gas fatalities are smaller, many resulting from gas line explosions. Wind and solar fatalities are not shown. They are only moderately higher than nuclear, all three being composed predominately of workplace accidents, most resulting in only a single fatality. As the numbers, in trillions of watt hours (TWh), make apparent, a switch from fossil fuels to nuclear and renewables for power is imperative to reduce the fatalities resulting from fossil fuel combustion as well as to fight climate change.

***

Is it possible that there could be another reactor accident resulting in a radiation release on the scale of Chernobyl or Fukushima? The answer must be that it is possible but highly improbable, given the improvements in reactor design for safety that have taken place in recent decades. Moreover, should such an event occur, it almost certainly would be in the aftermath of a natural disaster of unprecedented destruction, where like the Fukushima accident, the human toll, if any, caused by the radiation release would pale in comparison to the destruction and loss of life caused by the natural disaster itself, whether it be an earthquake, tsunami, volcanic eruption, meteor strike or something heretofore unimagined.

When considering such scenarios, an earthquake of unprecedented magnitude comes to mind. With the first tremor, reactors would be shut down. Thus, the worst that could be expected would be similar to the Fukushima radiation release. With an earthquake of that magnitude, however, the death toll from building collapse, fire, gas line explosions, chemical releases and other destruction directly attributable to the quake would far outweigh any health effects, such as the hypothetical increase in the cancer rate, that could be caused by radiation release from the ravaged nuclear power plants.

Russia's 2022 invasion of Ukraine has raised the question as to whether damage to a power reactor, such as those at the Zaporizhzhia Plant, from the attack by a hostile power could cause a major radiation release. Certainly, a power reactor would shut down automatically at the instant a missile hit its containment structure. Such structures are designed to withstand impact from aircraft crashes, and debris from wind and tornadoes. If a missile were powerful enough to penetrate the containment and the reactor pressure vessel, then a core meltdown and subsequent radiation release to the atmosphere would be possible.

An alternate scenario can be imagined that the reactor is occupied by a hostile army with considerable nuclear expertise. A destructive explosion could be set off within the containment at a place where the reactor is most vulnerable. On June 6, 2023, the Russian army is alleged to have used a similar strategy in setting off an explosion deep within the support structure of the Kakhovka Dam in Ukraine, causing it to rupture, and resulting in catastrophic flooding. Inside a reactor's containment, however, a steel pressure vessel encases the radioactive core, making it inaccessible to intruders. A powerful explosion outside that vessel might rupture piping that would cause the reactor core to melt. For radiation to be released to the atmosphere, the explosion would need to be so powerful as to also breach the containment structure. Whatever the damage to the plant, at Zaporizhzhia the radiation release to the atmosphere would be substantially smaller than what took place at Fukushima; since the reactors have been shut down for well over a year, the shorter half lived fission products—most importantly the radioactive iodine—have already decayed to insignificant levels.

Even with the worst accidents imaginable from nuclear reactors, health effects of radiation exposure would be small compared to the death and debilitation from accidents that happen in other industrial sectors. For a case in point, compare the 300 deaths of firefighters and other plant employees and the 15 thyroid cancer deaths following the Chernobyl disaster to the catastrophe that took place at the fertilizer plant at Bhopal, India.[26] In the early hours of December 3, 1984, the combination of a faulty valve and a leak in a large tank allowed water to mix with 40 tons of methyl isocyanine. A violent chemical reaction ensued, generating huge amounts of heat and pressure. Safety devices for neutralizing the toxic brew had been turned off. To prevent the tank from bursting, safety valves opened automatically, allowing the lethal plume to spew into the atmosphere over a city of 900,000.

In less than half an hour, the cloud engulfed the surrounding area. Nearly 4000 residents died in their beds, and streets were littered with human corpses and the carcasses of cows, dogs, and birds. Estimates of the death toll in the days that followed ran as high as 10,000, with 50,000 more suffering permanent disabilities. While Bhopal is the worst industrial accident to occur, there have been many others, some causing thousands of casualties. Consider just two accidents involving ammonia. A 1947 cargo ship fire in the Port of Texas City ignited 2300 tons of ammonium nitrate fertilizer. The resulting explosions killed nearly 600 people. More recently, on August 4, 2020, in Beirut, Lebanon, nearly 3000 tons of unsafely stored ammonium nitrate exploded. The blast, felt 150 miles away, killed more than 200 and injured 7000.

Accidents in numerous industries kill many more people each year than does radiation exposure. Likewise, chronic exposure to air pollutants from coal and other

---

[26] https://www.ncbi.nlm.nih.gov/pmc/articles/PMC1142333/

fossil fuels causes far more deaths and life-shortening than any radiation risks from nuclear power plants. Yet, too often, it is not objective statistics or careful radiation risk assessments that influence public opinion. Rather, the radiophobia of gut reactions induced by scary media coverage of every small nuclear mishap, no matter how trivial its public health consequences, has the larger impact on policy decisions. Given the ghostly history of radiation, radiophobia is understandable, but it should not be allowed to stymie the critical role that nuclear energy must play if global warming is to be brought under control.

# Chapter 14
# The Advance of Reactors

**Keywords** Breeder reactor history · Fast and thermal reactors · Uranium and thorium resources · Advanced reactors · Small modular reactors · Microreactors · Capital cost · Factory construction · Onsite construction · Incremental financing · Sodium and gas coolants · Molten salt reactors

After Enrico Fermi finished his experiments in 1942, the Chicago Pile-1 was moved from the densely populated neighborhood next to the University of Chicago to a more remote area on the city's outskirts. After the war ended, it became Argonne National Laboratory. Fermi encouraged his protégé, Walter Zinn, who became Laboratory Director, to initiate research into a kind of reactor very different from the water-cooled systems being developed by the navy.[1] Fermi was concerned with a problem for which a new type of reactor could be a solution. Supplies of uranium were limited, coming predominantly from the Belgium Congo. If nuclear energy was to become a major contributor to the world's electricity supply, its expansion could be limited in the decades to come by a paucity of uranium deposits. Concurrently, water- and graphite-moderated power reactors based on slow neutrons utilized only about 1% of the energy stored in uranium. They were powered primarily by uranium-235, which constitutes only 0.7% of natural uranium. The other 99.3% of uranium's energy was stored in uranium-238, whose energy made only a minor contribution to the power of commercial reactors utilizing graphite or water as moderator.

Thus came about the creation of the experimental breeder reactor, or EBR-I, followed later by EBR-II. These reactors were based on fission from fast neutrons, for in the megawatt (MeV) energy range, neutrons cause fission of both isotopes of uranium and of plutonium. Most importantly, of the neutrons absorbed by uranium-238, a fertile isotope, most create uranium-239, which undergoes radioactive decay to plutonium-239 within days. Plutonium-239 is fissile; neutrons over the entire energy range from MeV through eV cause plutonium to fission. If the reactors could

---

[1] https://www.ne.anl.gov/About/reactors/



E. E. Lewis, *Renewables or Nuclear*,
https://doi.org/10.1007/978-3-032-08074-5_14

breed more plutonium than the uranium they burned, the world's supply of reactor fuel would be multiplied by more than a factor of 100.

Water, however, could no longer be used as a coolant, for it was far too effective in slowing neutrons down to below the MeV range. In addition to being a fluid, the coolant needed to be composed of heavier material to inhibit neutron slowing down, but one that would not absorb many neutrons. Sodium satisfied both criteria. Designed at Argonne National Laboratory, EBR-I was the first reactor constructed in a remote area of Idaho, which became the national reactor test facility and decades later the Idaho National Laboratory.

The EBR-I reactor had a core of steel-clad uranium-235 cylindrical fuel rods, cooled by sodium and surrounded by a blanket of uranium-238. The neutrons escaping the core transformed the uranium-238 in the blanket to plutonium. The project's primary objective was accomplished, since EBR-1 created more plutonium than the uranium-235 it had consumed. It proved that reactors could be built that would greatly increase the world's supply of fissionable material and therefore of energy. The system was complete with a heat exchanger and turbine-generator, allowing small amounts of electricity to be produced. In December 1951, EBR-I produced what was arguably the first electricity from a nuclear reactor; it lit four 200 W lightbulbs, demonstrating reactors could be built that would both breed fuel and produce electricity.

The breeder reactor program continued with EBR-2, a larger reactor producing 20 megawatts (MW) of electricity. Over the 30 years of EBR-2 operation, many innovations increased efficiency and reliability as well as the ratio of bred plutonium to consumed uranium. Adjoining the reactor, the EBR program also created and operated facilities for processing spent fuel, separating the fission product waste, and recycling the bred plutonium back into the reactor. The years of successful operation also showed inherent advantages of sodium as a coolant. EBR-2 operated at higher temperatures than water-cooled reactors, and thus was more efficient in generating electricity. It also operated at much lower pressures than light water reactors (LWRs). Because these reactors were based on fast MeV neutrons, they became known as fast reactors, contrasting to reactors based on slow eV neutrons, which are called thermal reactors.

The concern with limited uranium resources to fuel the growing use of nuclear power for electricity motivated several nations to engage in breeder reactor development. In addition, the 1973 Middle East oil embargo caused France to build a fleet of pressurized water reactors over the following decade to provide its electricity and alleviate its dependence on imported oil. Fearing its future supplies of uranium would become limited, France simultaneously initiated a fast breeder reactor program and built the sodium-cooled Phoenix reactor. With similar motivations, Japan, the UK, and the Soviet Union also developed breeder reactors.[2] The next step in breeder technology in the USA took place in 1972, with a plan to build a sodium-cooled fast breeder reactor on Tennessee Valley Authority land next to the Clinch

---

[2] https://www.iaea.org/sites/default/files/gc/gc52inf-3-att6_en.pdf

River in Tennessee. The reactor was to be jointly financed by the federal government and the electric power industry; it would deliver 350 MW of electricity to the grid. However, the economic situation was changing and political support for the reactor was fading.

The perceived shortage of uranium in years immediately following World War II gradually dissipated. High prices had encouraged expanded uranium exploration, and large deposits of high-grade ore were found in several countries including Australia, Canada, Russia, Kazakhstan, Namibia, and the USA. Hence, securing uranium was accomplished without near the effort that had gone into exploiting petroleum reserves. Concomitantly, the growth of electricity demand slowed. As a result, the price of uranium fell, and the concern with shortages receded far into the future. Moreover, even though methods had been developed for reprocessing fuel from both water- and sodium-cooled reactors, their cost remained too high to compete with mining, processing, and enriching natural uranium. Then, following the accident at Three Mile Island, along with tightening reactor safety rules, President Carter passed legislation, halting the reprocessing of reactor fuel in the USA. This confluence of economic and political issues caused Congress to cancel the Clinch River project in 1983.[3]

Following the end of the cold war in late 1991, the international uranium market was flooded. Beginning in 1994 and lasting for two decades, the commercially financed government-industry partnership referred to as the Megatons to Megawatts™ operated. Over that time span, Russia blended 500 tons of highly enriched uranium—enough for 20,000 warheads—with natural uranium to produce 14,000 tons of low enrichment uranium suitable for fueling the water-cooled reactors widely deployed in the USA, France, and elsewhere.[4] Indeed, a substantial fraction of the uranium fuel for American power reactors during those decades came from diluted weapons material, bringing to mind "They shall beat their swords into ploughshares," from the Isaiah 2:4.

By the time the Megatons to Megawatts™ partnership ended, Russia had become part of western supply chains and a major source of uranium on international markets. Meanwhile, the capacity to process and enrich uranium in the USA greatly diminished, and the capability for breeding plutonium fuel in fast reactors was not pursued. Thus, with the onset of the Ukrainian War and NATO nations' attempts to boycott Russia's energy supplies, shortages are developing, and enrichment capabilities need to be restored rapidly in the West.

The French seem to be best positioned to take the lead in making their energy supply independent of Russia. They have maintained their plans to replace their fleet of LWRs over time with fast breeder reactors. An active research program toward that end is ongoing, and presently they are reprocessing LWR fuel, saving the depleted uranium for breeding in future fast reactors, and using the extracted

---

[3] https://en.wikipedia.org/wiki/Clinch_River_Breeder_Reactor_Project

[4] https://en.wikipedia.org/wiki/Megatons_to_Megawatts_Program#:~:ash-Ga

plutonium for refuel of their operating reactors, therefore reducing the amount of fresh low-enrichment uranium needed.

***

Over the last few years two overlapping nuclear developments have been receiving increased international attention for the contributions they may make toward fighting climate change: small modular reactors (SMRs), and advanced high-temperature reactors.[5] SMRs are defined as reactors with electric outputs 300 MW or less. Microreactors are a subset of SMRs with outputs of less than 50 MW. Across the globe there are more than 100 SMR designs at various stages of development[6]: 33 in North America, 1 in South America, 31 in Europe, 3 in Africa, and 28 in Asia.[7] Small modular reactors (SMRs) fall into two categories. The first evolve from currently operating power reactors, using low enrichment uranium fuel and high-pressure water as a coolant. The advanced reactors operate at higher temperatures, use fuel of about 20% enrichment, and are cooled by helium, sodium, or molten salts, with the sodium and salts being at low pressures. Current advanced reactor designs also tend to be about 300 MW or less.

Three advantages of SMRs stand out. First, except for common site infrastructure, raising capital to build one reactor module at a time is much less difficult than it would be for a gigawatt-sized reactor. Raising capital for succeeding modules would be spread over years, the frequency depending on the rate at which electricity and other energy demands grow. Second, the small size allows each reactor's modular core, its pressure vessel and much of the ancillary piping to be assembled in a factory. Factory manufacture is faster and subject to tighter quality control than can be accomplished with on-site construction. Such serial construction brought down the costs of ship building and aircraft construction, and it should reduce the costs of SMRs as well.[8] A third advantage is that these smaller reactors may be located on sites of shuttered coal plants of comparable power, where water is available and high-voltage transmission lines are in place.

In an advanced stage of development is the 462 MWe NuScale VOYGR-6, which employs pressurized water reactor technology.[9] It consists of 6 modules, each producing 77 MW, with shared site infrastructure. The plan is to follow with a 12-module 924 MW VOYGR-12 plant. Small 300 MW modular water-cooled reactors are also in the offing. Using established LWR technologies, the Westinghouse AP-300[10] and the Holtec SMR-300[11]are pressurized water reactors, while the GE-Hitachi BWRX-300

---

[5] Nuclear Energy Agency (2024). *Small Modular Reactors Dashboard: Second Edition.*

[6] The NEA Small Modular Reactor Dashboard_Third Edition.pdf.

[7] https://world-nuclear.org/information-library/nuclear-fuel-cycle/nuclear-power-reactors/small-nuclear-power-reactors

[8] Ibid.

[9] https://www.nuscalepower.com/-/media/nuscale/pdf/fact-sheets/smr-fact-sheet.pdf

[10] https://www.westinghousenuclear.com/energy-systems/ap300-smr

[11] https://holtecinternational.com/communications-and-outreach/smr/

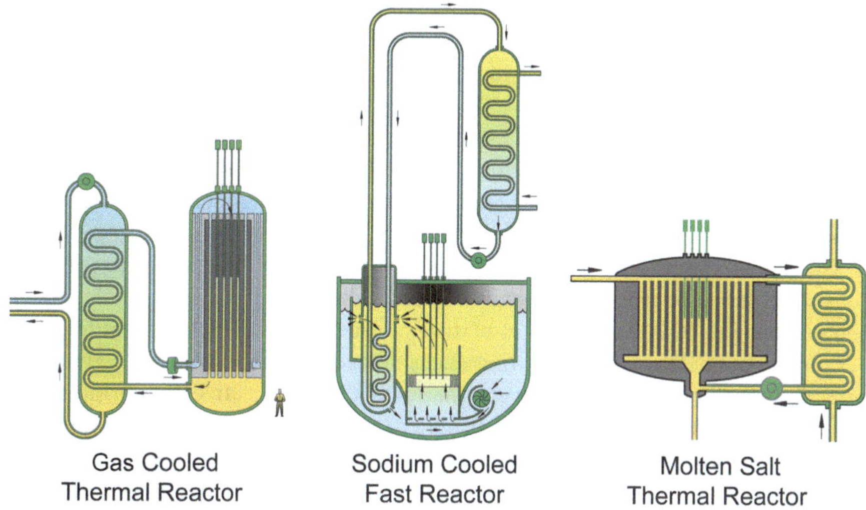



**Fig. 14.1** Advanced high temperature reactor concepts. (Source: DoE)

is a boiling water reactor.[12] All the new LWRs include passive safety systems relying on natural convection. Consequently, they contain no emergency pumps or valves that could fail if all emergency electricity was lost; their containments and other safety features are also improvements over twentieth century designs.

In the advanced reactor category, three concepts in the USA are presently at various stages of development and implementation in pilot plants. In addition to generating decarbonization electricity, they will produce heat at much higher temperatures than water-cooled reactors. They thus offer higher efficiency in electricity generation and attractive prospects for producing high-temperate heat for decarbonizing several industrial processes. Taken together, industrial processes account for nearly a quarter of the greenhouse gases emitted in the USA. These reactors are advanced from the water-cooled reactors that presently account for the lion's share of nuclear power production. None, however, are dependent on the success of basic research whose outcome is uncertain, such as in the case for longer term prospects of deep geothermal, ocean wave, and nuclear fusion energy.

In the USA, the three high-temperature advanced reactor concepts being pursued most actively are gas-cooled graphite-moderated thermal reactors, sodium-cooled fast reactors, and molten salt-cooled thermal reactors. Diagrams of each are shown in Fig. 14.1. The designation of fast or thermal is determined by whether the chain reactions are sustained by fast neutrons in the MeV range, or thermal neutrons in the eV range.

---

[12] https://nuclear.gepower.com/bwrx-300

The high-temperature gas-cooled Xe-100 reactor is being developed by X-Energy.[13] The graphite-moderated thermal reactor will consist of four 80 MW modules. It builds on experience with earlier high-temperature gas-cooled reactors. Two of these operated in the USA.[14] Their cores were graphite monoliths in which holes were drilled. Some holes were filled coated fuel particles, while others were coolant channels through which high-pressure helium carried the heat away. The Xe-100 modules also make use of graphite moderator, helium coolant, and coated particle fuel, but in pebble bed form.[15]

Diagramed on the left of Fig. 14.1 is one module of the Xe-100 reactor. Its core consists of a bed of tens of thousands of billiard ball-sized graphite pebbles. Each pebble contains 10,000 coated fuel particles. Called TRISO fuel, the particles are 0.5 mm in diameter and contain fuel encapsulated by three layers of carbon and ceramic-based materials that prevent the release of radioactive fission products. Helium flows downward through the pebble bed and transports heat out of the reactor to the heat exchanger, shown on the left of the reactor in Fig. 14.1. In the heat exchanger, very high-temperature steam is produced. The helium then recirculates through the pebble bed reactor core. The steam can be used as process heat or to produce electricity. Dow Chemical is partnering with X-Energy to build the first plant on the Gulf Coast at Seadrift, Texas, where its high-temperature steam will be employed to produce polyethylene resins, ethylene feedstocks, and other products.[16]

Sodium-cooled fast reactors, as explained earlier, have operated successfully for decades in France, Japan and other countries, as well as in the USA. Such projects have successfully bred more plutonium fuel than the uranium-235 burned and demonstrated how to reprocess the spent fuel for recycling in the same reactor. Although high temperature, sodium-cooled reactors operate at little above atmospheric pressure, thus eliminating high-pressure breaks as potential accident. Plans to construct two plants based on GE-Hitachi PRISM pool type reactor concept are underway: the 100 MW ARC-100 reactor in New Brunswick, Canada,[17] and the TerraPower 345 MW Natrium reactor at Kemmerer, Wyoming.[18]

Diagramed in the center of Fig. 14.1, the sodium-cooled Natrium reactor is being built under the auspices of the Department of Energy's Advanced Reactor Demonstration Program.[19] It is a pool type reactor, where the sodium coolant passes from the base of the reactor upward through the core picking up heat from a thousand or more metal-clad fuel rods. The sodium recirculates to the base of the reactor through a heat exchanger within the vessel, shown in Fig. 14.1 to the left of the

---

[13] https://x-energy.com/reactors/xe-100

[14] The Peachbottom-1 reactor in Pennsylvania and the Fort Saint Vrain reactor in Colorado.

[15] Earlier pebble bed experiments were in Germany: the AVR reactor, the experimental high-temperature reactor, and the 308 MW Thorium high-temperature reactor.

[16] https://x-energy.com/seadrift

[17] https://www.arc-cleantech.com/technology

[18] https://www.natriumpower.com/reactor-technology/

[19] Ibid.

reactor core. Sodium on the other side of the heat exchanger carries the heat outside the reactor vessel to a second heat exchanger shown in Fig. 14.1 above the vessel. There the heat is transferred across the heat exchanger from sodium to molten salt.

A pioneering feature of the Natrium system borrows an idea from the concentrated solar system discussed in Chap. 9. The nuclear system combines the reactor with thermal energy storage: Molten salt is heated as it circulates through a second heat exchanger shown above the vessel, and then stored in a large well-insulated tank, called the hot tank. Then, as needed, heat is transferred from the molten salt to boil water, which then powers a steam turbine to generate electricity. Thus cooled, the molten salt is stored in a second large tank, called the cold tank, until entering the heat exchanger above the vessel to be reheated by hot sodium from the reactor vessel.

The two-tank storage system buffers the reactor from the turbine-generator. Consequently, while the reactor operates steady state at 345 MW, the turbine generator can produce up to 500 MW of power for more than 5 h to take up the slack when the wind isn't blowing, or the sun isn't shining. Conversely, the system's output can be reduced to satisfy diminished demand that may be a fraction of the 345 MW at which the reactor operates.

Reactors cooled by molten salt are also receiving increased commitment. They build on work at Oak Ridge National Laboratory in the 1960s. With support from the Tennessee Valley Authority and other partners, Kairos Power is developing the KP-140, a 140 MW high-temperature thermal pebble bed reactor, moderated by graphite, and cooled by fluoride molten salt.[20] The reactor will employ ceramic-coated fuel pebbles that are nearly identical to those used in the high-temperature gas-cooled Xe-100 reactor. Using molten salt instead of a gas coolant, however, should allow the reactor to operate at the same high temperatures as gas-cooled reactors, but at only a little above atmospheric pressure. It is at an earlier stage of development than the other concepts, with the need to first build a small test reactor at Oak Ridge before proceeding to the construction of a KP-140 plant.

Taking molten salt concepts further will be Terrestrial Energy's IMSR reactor. It will be a fluid-fueled reactor in which the coated fuel particles are dissolved in the molten fluoride salt which also acts as the coolant.[21] An experimental reactor of this type was operated in the 1960s at Oak Ridge National Laboratory.[22] Diagrammed on the right of Fig. 14.1, the fluid fuel of molten salt passes upward through a core of graphite rods and exits to the heat exchanger to the right of the reactor before circulating back to the base of the reactor. The heat is carried to a second heat exchanger, not shown in the diagram, where high-temperature steam is produced.

A collaborative molten salt project led by TerraPower and the Southern Company has also been initiated. The two companies are examining the possibility of using

---

[20] https://www.nei.org/news/2021/hermes-kairos-power-new-path-advanced-reactors

[21] https://www.terrestrialenergy.com/advanced-nuclear-reactor-design/

[22] https://www.ornl.gov/molten-salt-reactor/history

molten chloride salt in a fluid-fueled fast breeder reactor.[23] Unlike sodium-cooled fast breeder reactors, which breed plutonium-239 from uranium-238, it would breed uranium-233 from thorium-232. While uranium deposits are sufficiently plentiful to fuel breeder reactors for centuries, thorium deposits worldwide are estimated to three times as large as uranium deposits.

Microreactors producing less than 50 MW of electric power are also under development in the USA and abroad.[24] They differ significantly from their larger cousins, the SMRs that produce substantially more than 50 MW. Microreactors are manufactured entirely within factories and fit in shipping containers to be delivered by road, rail, sea, or even by air to where they are needed. They also have much longer fuel lives than other SMRs, some of them designed to operating up to 20 years before refueling. Moreover, refueling in not done on site. Instead, microreactors are returned in their containers to where they were manufactured. There, spent fuel is removed and stored, and the reactors refurbished for shipping to their next assignments. Concomitantly, identical reactors may be shipped to sites to replace the originals immediately following their removal.

Among microreactors under development is the Xe-Mobile, which is a smaller form of the Xe-100 high-temperature gas-cooled reactor. The Westinghouse eVinci is a thermal reactor using TRISO fuel in a graphite-moderated core, with heat pipes for cooling. The Oklo Inc. Aurora is a fast reactor with metal fuel, also with a heat pipe cooling system. The Last Energy reactor uses standard pressurized water technology, but on microreactor scale of 20 MW to allow full factory construction and streamlined licensing. In the USA, the Department of Defense is backing microreactor development for use in remote locations or where emergency power is needed. Generally, they are air cooled, allowing them to be sited where no water is available. Microreactors, or pairs of them, may also offer a path to decarbonizing merchant shipping, particularly large container and tanker ships.[25]

The foregoing advanced concepts operate at temperatures higher than the 300 °C (572 °F) temperatures of water-cooled reactors. Consequently, while water-cooled reactors may be used to supply heat for desalination, district heating, and some other industrial processes, the anticipated ranges between 350 and 650 °C (660 and 1200 °F) of the advanced reactors allows replacing fossil fuels with nuclear heat for petrochemical processing, ammonia synthesis, and high-temperature electrolysis for producing hydrogen. Clean hydrogen thus produced may then be used to obtain the even higher temperatures required to make steel and concrete. If such advanced reactors can be deployed soon enough, they should be capable of generating much of the carbon-free power needed for industrial processes as fossil fuel plants are shuttered.

✳✳✳

---

[23] https://world-nuclear.org/information-library/current-and-future-generation/molten-salt-reactors

[24] https://www.energy.gov/ne/articles/what-nuclear-microreactor

[25] https://spectrum.ieee.org/nuclear-powered-cargo-ship

Novel reactor design will be critical in supplying dispatchable carbon-free power that can satisfy the stringent load-following and fast-ramping peaking capabilities needed to complement intermittent wind and solar generation as its share of grid capacity grows. Such reactors must be able to operate at very low powers for much of the time yet be capable of ramping up and down in power very rapidly to supply erratic residual demand profiles, such as shown in Figs. 5.4 and 5.5. The high-temperature gas-cooled Xe-100 reactor is designed to have such peaking capabilities, as is the sodium-cooled Natrium reactor when combined with molten salt energy storage. With load following capabilities that surpass those of large present-day reactors, microreactors may also be placed near large wind or solar farms to complement VRE intermittency.

Future water-cooled reactors may draw on the experience of naval propulsion reactors to design the carbon-free peaking plants. Naval reactors must operate at a small fraction of full power for long periods of time—for example, when a ship is cruising, or at dockside in places where auxiliary electricity is inadequate. Critically, such propulsion reactors must ramp to full power in minutes or less when required by military situations and then drop quickly back to low level operation. Space and weight are not limited on land as they are on shipboard, and naval reactor fuel must last 30–40 years, while that in power reactors is removed after 5 or 6 years of exposure. Thus, the challenges and costs of building highly maneuverable reactors should be less daunting for electric power production than for the propulsion of warships.

# Chapter 15
# Waste or Resource

**Keywords** Nuclear fuel reprocessing · Waste disposal · Weapons proliferation · Coal ash · High level waste · Low level waste · Spent fuel · Deep geological repositories · Weapons grade plutonium · Enriched uranium · Dirty bombs

On December 22, 2008, a toxic waste spill of monumental proportions sullied 300 acres adjoining the Tennessee Valley Authority Kingston Coal Plant, located on Watts Bar Lake near Harriman, Tennessee.[1] A 60-foot-high earthen dike constraining a large pond filled with a slurry of coal ash and water suddenly gave way. A wave of more than a billion gallons of sludge descended on the surrounding landscape, covering it with as much as 6 feet of the slurry. As pictured in Fig. 15.1, the wave's ferocity damaged dozens of houses, downed trees, destroyed power lines, washed out a road, broke water mains, and obstructed a rail line. The slurry contaminated more than 300 acres and caused fish kills by contaminating the Emory and Clinch Rivers, both tributaries of the Tennessee River.

The dyke failure resulted in millions of dollars of property damage, but thankfully there were no fatalities that day. The health risks were more insidious, stretched over the decade that followed, attributable in large part to the mercury, lead, arsenic, and other toxic materials contained in coal ash. Death or disability came later to the nearly 1000 workers who spent the next decade in the billion-dollar cleanup of the site. Thirty-six succumbed to leukemia, brain cancer, lung cancer, or other diseases closely associated with toxic chemicals. Others were in poor health, needing respirators to breathe. A lawsuit was brought by more than 800 of those harmed by the spill and resulted in a \$27.8 million payment to the plaintiffs after 5 years of litigation.[2]

The coal ash waste disposal site located next to the Kingston plant is only one of the more than a thousand scattered across the USA. There are ponds, limestone quarries, and other confinements, many that are susceptible to leaching of known

---

[1] https://www.epa.gov/tn/epa-response-kingston-tva-coal-ash-spill

[2] https://www.nrdc.org/stories/hundreds-workers-who-cleaned-countrys-worst-coal-ash-spill-are-now-sick-and-dying



E. E. Lewis, *Renewables or Nuclear*, https://doi.org/10.1007/978-3-032-08074-5_15

**Fig. 15.1**  Kingston coal slurry spill. (Source: TVA)

carcinogens into rivers, wells, and other groundwater supplies. The pollution persists without attention-grabbing failures such as that at Kingston.[3] Nor is coal the only fossil fuel creating waste and pollution problems. There are thousands of orphaned oil and gas wells—nearly 150,000 in Texas alone[4]—many which are leaking methane into the atmosphere. While it does not remain in the atmosphere for more than a decade or two, as a greenhouse gas, methane is as much as 80 times more powerful than carbon dioxide. A federal program that subsidizes capping the wells does exist, but the issue is urgent, and a program needs to be put in place with stiff, easily enforced penalties to greatly accelerate the capping process. This is low-hanging fruit for reducing greenhouse gas emissions. Over the short term, emission reductions can be accomplished much faster by capping wells than by building solar or wind farms or by the construction of nuclear reactors.

If we compare the waste produced by a coal plant with a nuclear power reactor producing the same amount of electricity, say 1 gigawatt, two differences immediately stand out. First, critics often pointed out that the radioactivity of high-level nuclear waste lasts a very long time, a small part of it is still significantly radioactive for tens of thousands of years. Most of that nuclear waste, however, decays away in a few hundred year and the remainder can be recycled, topics to which we shall return. In contrast, chemical carcinogens found in coal ash don't decay away; they remain in the environment forever.

Second, the high-level nuclear waste in actuality consists of bundles of used fuel rods—referred to as "spent fuel"—removed from reactors. The amount of it is minuscule compared to the waste from a coal plant with the same power rating. Each year, a coal plant operating at 1 gigawatt (GW) produces roughly 200,000 tons

---

[3] https://earthjustice.org/feature/coal-ash-map-sites-legacy-inactive-regulated

[4] Houston Chronical Editorial, "Take the money and Plug. Orphan wells mess in Texas," January 12, 2023.

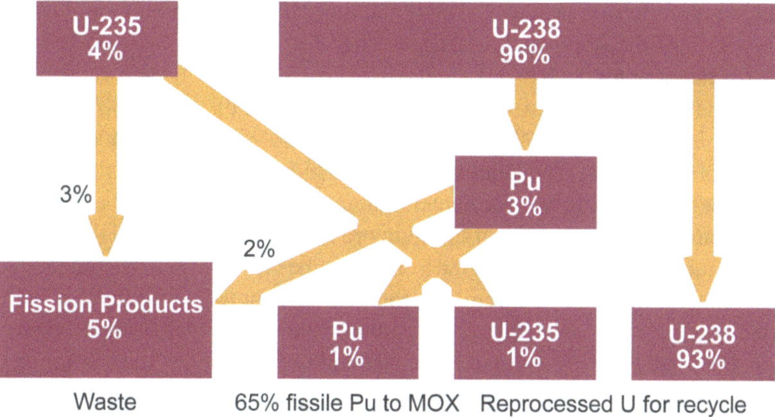



**Fig. 15.2** Composition of new and spent fuel from water-cooled reactors. (Source: World Nuclear Association)

of waste ash and spews about six million tons of $CO_2$ into the atmosphere.[5] In contrast, the gigawatt (GW) reactor produces only about 25 tons of spent fuel in a year. Therefore, the ratio of ash to spent fuel mass is 8000 to one, without counting the $CO_2$. Consequently, a GW coal plant must dispose of close to 20 large railroad cars of ash per day, while only about a third of the fuel in a reactor needs to be replaced and stored at 18–24 months intervals.

***

Controversy over what to do about nuclear waste centers on high-level as opposed to low-level waste. Low-level waste consists of such things as contaminated clothing, paper, rags and tools that contain small amounts of mostly short-lived radioactivity. The waste is generated at hospitals, research laboratories and other institutions as well as at nuclear reactors. Low-level waste does not require shielding during handling and transport, and it is disposed of safely at shallow burial sites. Low-level waste comprises 90% of the volume but only 1% of the radioactivity of all nuclear waste.[6] High-level nuclear waste is what causes outsized public concern. The spent fuel from nuclear reactors is what constitutes high-level waste. It, however, can be reprocessed, converting most of it to a valuable energy resource.

As indicated in Fig. 15.2, fuel from present-day power reactors typically consists of uranium enriched to about 4% uranium-235 and 96% uranium-238. As the figure indicates, about three-fourth of the uranium-235 undergoes fission, hence creating highly radioactive fission products (i.e., the fragments that result from the splitting uranium atoms.). The other one-fourth remains in the spent fuel. A little more than 3% of the uranium-238 absorbs a neutron and then undergoes radioactive decay to produce plutonium (Pu). About two-third of that plutonium undergoes fission,

---

[5] https://www.gem.wiki/Estimating_carbon_dioxide_emissions_from_coal_plants

[6] https://world-nuclear.org/nuclear-essentials/what-is-nuclear-waste-and-what-do-we-do-with-it

producing more fission products. The remaining plutonium and uranium constitute roughly 95% of the spent fuel, and as we shall see, it can be recycled and used as fuel for future reactors.

Overwhelmingly, the fission product isotopes have half-lives of 30 years or less. In ten half-lives, or about 300 years, the radioactivity of fission products is reduced to negligible amounts. The plutonium is suitable fuel to be recycled in today's water-cooled reactors. A use for depleted uranium will emerge in the more distant future when the richer deposits of uranium ore have been mined, and the production of reactor fuel from much lower-grade ores becomes more expensive than recycling spent fuel. The depleted uranium can then be transmuted to plutonium in future breeder reactors and subsequently fissioned. Other than plutonium, there will remain trace amounts of radioactive transuranic (i.e., heavier than uranium) elements, with half-lives of thousands of years. They include neptunium, americium, and curium. These, however, can be stored and then depleted by placing them in breeder reactors.

France, where nearly three quarters of the electricity is generated by nuclear power plants, has an advanced program for recycling spent fuel.[7] At a plant in La Hague, the highly radioactive fission products are separated from the rest of the spent fuel. A vitrification process is used to imbed the fission products in a glass that has stable chemical properties, and the glass is encapsulated in stainless steel containers. The containers need to be isolated for only a few hundred years. By then the fission products will have decayed to the level of less than background radiation. The plutonium is separated and combined with newly refined uranium to form mixed oxide fuel (MOX), which is then used for refueling existing reactors. By separating the fission products and plutonium, the radiotoxicity of the remaining uranium is reduced by 80%. That uranium is placed in interim storage where it can be retrieved and blended in fuel for future reactors.

Finland, Sweden, France, Canada, Switzerland, Japan, and other industrialized nations are at various stages of planning deep geological repositories for the final disposal of spent fuel from their reactors. These repositories are designed to receive fuel over many decades and to isolate and contain the radiological material for hundreds of thousands of years. The repositories must not rely on human maintenance or intervention. Even while France is recycling fuel, it is also keeping its options open by making plans for a deep geological repository where nuclear waste can be stored 1500 feet below ground in a stable environment, embedded in impermeable claystone.

Finland's Onkalo spent fuel repository, which is being built in accordance with governmental objectives set 40 years ago, is complete and will start receiving spent fuel shortly.[8] It is located more than 1200 feet beneath the surface in bedrock that has been stable for two billion years. The repository, located near Finland's three

---

[7] https://www.iaea.org/newscenter/news/frances-efficiency-in-the-nuclear-fuel-cycle-what-can-oui-learn

[8] https://group.vattenfall.com/press-and-media/newsroom/2023/finland-to-open-the-worlds-first-final-repository-for-spent-nuclear-fuel

reactors, will eventually have 100 tunnels, totaling 40 miles in all. The nuclear fuel will be encased in double layer metal canisters and placed in slot holes bored in the floors of the tunnels and barricaded with concrete. The facility should have the capacity to isolate a century's worth of spent fuel, incased in well over 3000 canisters. The entire facility will then be sealed and inaccessible.

An ongoing debate revolves around whether it is wise to sequester spent fuel in deep depositories in such a way that it cannot be retrieved. At some time in the distant future, uranium ore deposits may become sufficiently scarce that it is less expensive to reprocess spent fuel and use its uranium-238 in breeder reactors to increase the world's energy supply by a 100-fold. Moreover, the world's thorium-232 deposits are much larger than those of uranium, and like uranium-238, if it absorbs a neutron, it undergoes radioactive decay to become uranium-233, a fissionable isotope, but one that does not occur naturally. Consequently, breeder reactors may be built on either a uranium-238/plutonium-239 or a thorium-232/uranium-233 cycle. France has been among the most aggressive of several nations in developing reprocessing and in designing uranium-based breeder reactors to decrease its reliance on imported uranium. Breeder reactors based on thorium are at an earlier stage of development.

In shorter term, the expense of reprocessing may be substantially reduced relative to that of producing fuel by enriching newly mined uranium by commercializing uses of the separated fission products. With half-lives of about 30 years, strontium-90 and cesium-137 account for nearly all the radioactivity separated from the uranium and plutonium in reprocessing. Both have numerous medial applications. In addition, cesium-137 is used for food sterilization, and strontium-90 may serve as the basis for Radioisotope Power Systems. Several industrial concerns are exploring a wide range of their uses.[9] These include providing heat and electricity for future space missions and deep-sea mining, as well as powering remote outposts and providing transportable emergency generators.

***

The situation in the USA regarding nuclear waste is complicated because the waste is generated by both military and commercial sources. Thus, waste disposal has become entwined with nuclear weapons' proliferation issues and immersed in domestic political disputes. Waste from nuclear weapons production, naval reactors and other military uses is handled separately from the spent fuel of civilian power reactors. The long-lived military waste is sequestered at the Waste Isolation Pilot Plant located 26 miles from Carlsbad, New Mexico. It is permanently isolated more than 2000 feet beneath the surface in an ancient salt formation. What to do about the waste from commercial power reactors has a more convoluted history.

In the 1970s, France, the UK, Japan, and Russia operated nuclear fuel reprocessing centers. Three were planned in the USA, and one, located near Buffalo New York,

---

[9] https://www.ans.org/news/2025-06-10/article-7046/nuclear-fuel-cycle-reimagined-powering-the-next-frontiers-from-nuclear-waste/

operated from 1966 to 1972.[10] Another located at Morris, Illinois, operated only as a pilot plant and was closed in 1974. The construction of a third, at Barnwell, South Carolina, was aborted in 1977 as the result of a change in government policy.

Politically, the concern was that separating plutonium from spent fuel would increase the threat of nuclear weapons proliferation. The logic was that with the plutonium intermingled with the highly radioactive fission products in the fuel, it was secure, but once separated from the fission products by reprocessing, it would be more susceptible to theft. In 1977, President Carter banned the commercial reprocessing of fuel. In 1981, President Regan lifted the ban, but then a year later Congress passed the Nuclear Waste Policy Act which required the federal government to take possession of spent fuel and permanently dispose of it. Thus, there would be no repossessing.

The federal government began taxing reactor owners in the 1980s, collecting the 44 billion dollars it had earmarked for a permanent nuclear waste facility. In 1994, work was initiated to isolate spent fuel from power reactors in tunnels about 1000 feet under Yucca Mountain. The remote location is adjacent to where nuclear weapons had been tested in the Nevada desert, about 100 miles from Los Vegas. Although the project was technically feasible, a long-lasting political impasse between federal and state government, caused the project to be cancelled in 2010.[11]

There is no political agreement as to where or in what form high-level waste should be stored permanently in the USA. At present, it is left in cooling pools within the reactor containment until much of the decay heat produced by the fission products has dissipated. It is then placed in dry storage casks such as shown in Fig. 15.3, most often located at the reactor site. For our GW reactor example, only a little more than one cask per year of reactor operation is needed. The casks are designed and constructed to withstand, without leaking, earthquakes, tornadoes, fire, flooding, and even an aircraft crash.

***

The question arises: If the Nuclear Waste Policy Act were amended to allow reprocessing of US spent fuel, would it be recycled as it is in France? For two reasons, it seems less likely. First, France has a national power utility that answers to the government. As such, it can plan for the long term without needing to show a profit on a year-to-year basis. Most US reactors are privately owned by regulated monopolies or by merchant generating companies who compete in wholesale electricity markets. In either case, economics determines where they buy their fuel. Present prices for uranium ore make it more economical to purchase fuel fabricated from newly mined uranium than to purchase mixed oxide fuel fabricated from recycled plutonium and uranium. Second, among the citizenry, worry persists that plutonium separated during reprocessing would be more susceptible to theft. Such theft

---

[10] https://world-nuclear.org/information-library/nuclear-fuel-cycle/fuel-recycling/processing-of-used-nuclear-fuel.aspx

[11] https://slate.com/technology/2013/01/nuclear-waste-storage-why-did-yucca-mountain-fail-and-what-next.html

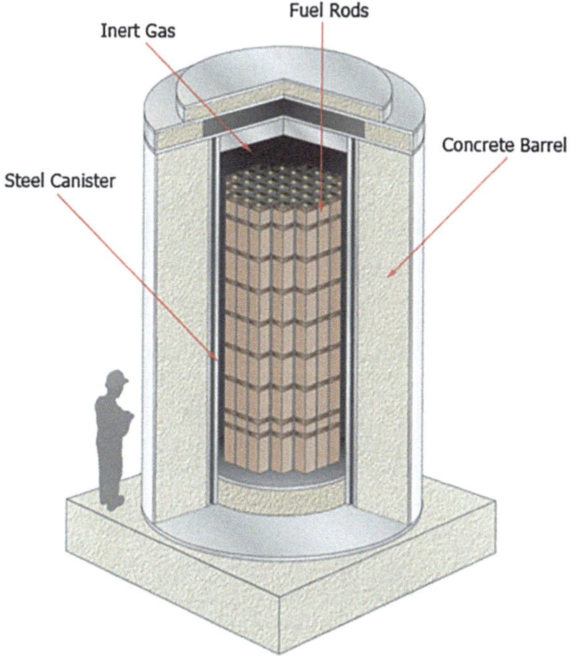


**Fig. 15.3** Dry spent fuel cask. (Source: NRC)

would accelerate nuclear weapons proliferation. But this is not the case, as may be understood by taking a closer look at plutonium and the weapons that utilize it.[12]

In nuclear reactors, plutonium is produced by uranium-238 absorbing a neutron. The resulting uranium-239 then undergoes radioactive decay to become plutonium-239. Figure 15.4 shows the growth of the two most prevalent plutonium isotopes over 3 years, the time at which fuel remains in a power reactor. Plutonium-239 builds up rapidly at first, but then the curve flattens. By then, roughly one-third of the power in present-day water-cooled reactors results from fission of plutonium-239. Plutonium-240 accumulates more slowly, for it is produced by Plutonium-239 capturing a neutron. Unlike plutonium-239, however, plutonium-240 undergoes spontaneous fission, continuing to emit neutrons long after the reactor is shut down.

Plutonium-240 is a problem for designers of nuclear weapons, for the neutrons emitted from spontaneous fission will cause the weapon to preignite, and the higher the percentage of plutonium-240, the higher the probability for preignition. Preignition results in what is sometimes called a fizzle; it yields only a tiny fraction—if any—of the explosive energy that the weapon was intended to produce. To be classified as weapons grade, plutonium is required to contain less than 8%

---

[12] https://world-nuclear.org/information-library/plutonium.aspx

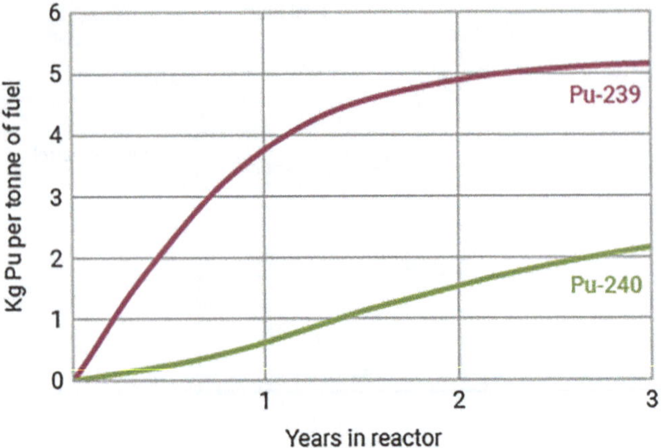



**Fig. 15.4** Plutonium buildup in a power reactor. (Source: World Nuclear Association)

plutonium-240. Recycled plutonium must be classified as reactor grade, which has more than 19% of the 240 isotope.[13]

To produce weapons-grade plutonium, uranium fuel can be in a reactor for no more than 3 months. Thus, nations attempting to obtain plutonium-based weapons build reactors for irradiating fuel intensively for short periods of time, and inspectors can easily distinguish them from reactors designed to produce only power for electricity or propulsion. Consequently, reactors designed for power production cannot be used to produce plutonium for nuclear weapons. Therefore, reprocessing power reactor fuel would not result in nuclear weapons proliferation.

These statements remain true for fast breeder reactors based on the uranium–plutonium cycle, for plutonium-240 isotope will be produced, making the spent fuel unsuitable as a weapons material. Recently, interest has been growing in building breeder reactors based on a thorium–uranium-233 cycle. Thorium-232, analogous to uranium-238, is a fertile material; only very high energy neutrons will cause it to fission. However, if it captures a neutron, it undergoes radioactive decay and produces uranium-233, which like uranium-235 is fissionable. The first generation of thorium uranium-233 breeder reactors, however, must be fueled with uranium-235 or plutonium since uranium-233 doesn't occur naturally. Thereafter, some of the bred uranium-233 could be used to fuel the next cycle.

Thorium had garnered interest because it is so plentiful. Some estimates are that there is three times as much thorium that could be mined as uranium. Moreover, its deposits are spread more widely over the earth's crust. India, in particular, has had ongoing research on thorium fueled reactors because it has widespread thorium resources, but no uranium. Thus, while worldwide uranium deposits should be

---

[13] https://world-nuclear.org/information-library/nuclear-fuel-cycle/fuel-recycling/plutonium

adequate for hundreds of years—even without breeder reactors—security issues often cause nations to favor materials for which they have plentiful domestic supplies.

Like plutonium-239, pure uranium-233 is a weapons-grade material. However, along with the uranium-233, small amounts of uranium-232 are produced, and it is not weapons grade—quite the opposite. While the mechanism is more complicated than that of plutonium-240,[14] the presence of even a small amount of uranium-232, even 0.005%, bars uranium-233 from being used for weapons.[15] Moreover, like the buildup of plutonium-240 shown in Fig. 15.4, the ratio of uranium-232 to uranium-233 increases with time. Analogous to the uranium-plutonium cycle, in a thorium-uranium-233 cycle, the fuel could be left in the reactor for only a very short period of time if it were to be used to produce weapons material, and the reactor's design would need to be radically different from those of breeder or other types of power reactors. Likewise, inspectors would have no difficulty in telling the difference.

None of the above is to say that nuclear weapons proliferation in not a concern, but only that reactors built solely to produce electricity are not a source of weapons-grade materials. Proliferation is prevented through adherence to the Nuclear Nonproliferation Treaty (NPT), which went into effect in 1970. It was signed by 191 nations, including the five counties that had nuclear weapons at that time: the USA, Russia, the UK, France, and China. Adherence to the treaty is monitored by the International Atomic Energy Agency (IAEA). Among its duties are preforming inspections, and monitoring inventories of nuclear materials. IAEA inspectors could easily determine whether a reactor was designed and configured to produce weapons-grade materials.

None of the four countries that have obtained nuclear weapons since 1970 are NPT signatories, and none have obtained materials for their weapons from reactors designed solely for electric power production. The European educated A. Q. Khan returned to Pakistan after stealing blueprints for a high-speed gas centrifuge used to enrich uranium for civilian purposes. He thus enabled Pakistan to build nuclear weapons based on highly enriched uranium. To make matters worse, he then sold or traded the weapons technology to North Korea, Iran, and Libya.[16] India modified a research reactor provided by Canada as part of the Atoms for Peace project to produce weapons-grade plutonium.[17] Similarly, with French assistance, Israel built a research reactor, which many believe was later modified to produce weapons-grade plutonium.[18] North Korea obtained its nuclear technology from Russia when the two were close allies before the NPT went into effect. If IAEA inspectors had been

---

[14] *Langford, R. Everett (2004). Introduction to Weapons of Mass Destruction: Radiological, Chemical, and Biological. John Wiley & Sons. p. 85.*

[15] https://nuclearweaponarchive.org

[16] https://ahf.nuclearmuseum.org/ahf/profile/q-khan/

[17] https://nsarchive.gwu.edu/briefing-book/nuclear-vault/2022-12-09/us-canada-and-indian-nuclear-program-1968-1974

[18] https://armscontrolcenter.org/wp-content/uploads/2020/03/Israel.pdf

allowed to examine these nations' nuclear facilities, they could have easily identified the centrifuges or other equipment for obtaining high enrichment uranium or the reactors built or modified to obtain weapons-grade plutonium.

Understandably, there is nevertheless an ongoing concern that terrorists could gain access to the necessary nuclear technology and wreak widespread devastation. A recent National Academies study[19] argues that since nuclear weapons, materials, and expertise are controlled almost entirely by state actors (i.e., nations), the ability of terrorists to carry out a nuclear attack would require either the complicity of a state with nuclear weapons, or the failure of such a state and of its controls. Such chaotic failure could be caused by economic collapse, civil war or overthrow of the government. The 1991 collapse of the Soviet Union, for example, presented the risk that terrorists could gain access to nuclear martials and possibly even weapons in the chaos that followed.

The National Academies study[20] distinguishes between nuclear weapons and improvised nuclear devices. Theft of an intact weapon would require needing only the capability of transporting and igniting it. The theft of sufficient highly enriched uranium or bomb-grade plutonium may be the largest impediment to improvising a clandestine weapon. Without expertise and the well-equipped laboratory of a nation state, however, fashioning a workable nuclear device would be highly problematical. An improvised device would be of questionable reliability, its yield difficult to predict, and its construction and transport fraught with consequences of accidental detonation.

The salient conclusion, however, must be that building reactors solely to produce electricity or heat for industrial use does not add to the terrorists risk already present without them. Even the most advanced power reactors employ enrichments of no more than about 20%, while bomb-grade uranium must be enriched to about 90%. The exception is for the higher enrichments needed to reduce the weight and volume of naval propulsion reactions. Such uranium is held under the tight control of the defense establishments of the nations with nuclear navies. As explained in the preceding section, the plutonium produced in electric power reactors is degraded by inseparable plutonium-240, and therefore, it is not a proliferation risk. Likewise, the proliferation risk of future thorium cycle reactors is nullified by the inseparable production of small amounts uranium-232 along with the uranium-233.

Dirty bombs present a quite different threat that is more likely but far less severe than the detonation of a nuclear weapon. Terrorists would use a chemical explosion or other mechanism to trigger the spread a dangerous amount of radioactive material across the surrounding environment.[21] The primary method of prevention is to block terrorists from obtaining necessary building material from the many large radioactive sources that are widely used for medical imaging, cancer treatment,

---

[19] National Academies (2024). *Nuclear Terrorism: Assessment of U.S. Strategies to Prevent, counter, and Respond to Weapons of Mass Destruction*.

[20] Ibid.

[21] Ibid.

sterilization, food processing, industrial scanning, and more. These must be protected from theft and at the end of useful life securely stored.

In assembling a massive amount of radioactive material in a compact device, the perpetrators would face inevitable danger of subjecting themselves to acute radiation sickness or death. Conversely, the heavy weight of shielding needed to protect the perpetrators from deadly levels of radiation exposure would make the device near impossible to conceal and move into place. Moreover, the primary effects a dirty bomb would no doubt be economic dislocation and widespread panic from the public's ingrained fear of radiation rather than harm from radiation exposure. A combination of rapid evacuation followed by decontamination or of sheltering in place would greatly limit exposure. Abandonment of the affected area until cleanup of the affected environment is completed, however, would be expensive and possibly long lasting.

# Chapter 16
# The Dismal Science of Grids

**Keywords** Cost of electricity · Levelized costs · Fixed and variable costs · Renewable energy · Nuclear energy · Direct and system costs · Grid level costs · Energy markets · Vertically integrated utilities · Plant lifetimes · Capacity factors · Missing money problem

Examining a monthly electric bill can be an instructive exercise, for it is where many of the considerations of the preceding pages are reduced to their cost. The format in which the information is presented may vary, depending on the state in which you live and the distribution utility that serves you. Generally, in the USA, the bill divides the costs, measured in dollars per megawatt hour ($/MWh), into generation, transmission, and distribution, corresponding to the fundamental parts of an electric grid. On the bill for my condominium, shown as in Table 16.1, generation is 54.02 $/MWh, transmission 14.70 $/MWh, and distribution 19.98 $/MWh. Generation about which much of our discussion has been centered is the largest expense. Taxes and services are added to the bill. They are independent of the amount of electricity used, but if we add them in dollars and divide by the 0.724 MWh consumed, we obtain an equivalence of 27.40 $/MWh. On my bill, these charges are roughly divided between state and local taxes and the distribution utility's fixed costs.

In Table 16.1, my electricity costs are totaled and the percentages listed. Two state subsidies are included. The 5.02 $/MWh, titled Renewable Portfolio Standard, accounts for the certificates described in Chap. 3 that the distribution utility must buy at market value from wind and solar generators to meet the state's requirement for the percentage of electricity that must be obtained from renewable sources. The 1.95 $/MWh, titled Zero Emission Standard, is the Illinois subsidy to assure that the state's carbon-free nuclear power plants remain economically viable while competing with low natural gas prices and subsidized variable renewable energy (VRE) in the deregulated electricity markets.

Not appearing on electric bills are federal subsidies discussed in Chap. 3. They are in the form of investment and production tax credits. The investment tax credits



E. E. Lewis, *Renewables or Nuclear*,
https://doi.org/10.1007/978-3-032-08074-5_16

**Table 16.1** Cost of
electricity

| Cost | $/MWh | % |
| --- | --- | --- |
| Generation | 54.02 | 43.9 |
| Transmission | 14.70 | 11.9 |
| Distribution | 19.98 | 16.2 |
| Taxes and services | 27.40 | 22.3 |
| Renewable portfolio st. | 5.02 | 4.10 |
| Zero emission st. | 1.95 | 1.95 |
| Total | 123.07 | 100 |

effectively reduce the cost of building decarbonized power plants, while the production tax credit reduces the cost of operating them. Since VRE has no fuel cost, the production tax credit reduces wind and solar operating cost to almost zero, and at times, it is negative. In effect, VRE can give away energy at times and still make money, for its production costs are more than covered by the combination of federal production tax credits and the sale of state-issued renewable portfolio certificates.

***

The economics of costs, geography, and subsidies play interacting roles in determining what sources of electricity will be most important in reaching the 2050 goal of carbon neutrality. Some energy sources are still in the form of research and laboratory experiments, and for these, viability has yet to be demonstrated. Thus, at this point, they cannot be counted upon to contribute significantly to the fight against climate change. Harnessing the energy of deep geothermal heat, ocean waves, and nuclear fusion fall into this category. Other energy sources, such as hydroelectric and geothermal power are well established and currently supply power to electric grids. Their future expansion, however, is severely limited by geography in the USA and many other countries. Hence, the following discussion focuses on the economics of the three most prominent sources of carbon-free electricity, wind, solar, and nuclear power, and on the fossil fuels they must replace.

The numbers on an electric bill are for the most part based on components of the levelized costs of electricity (LCOE), which is defined as the dollars per megawatt hour that must be charged to recover the costs of building a plant and operating it over a specified cost recovery period, that is, its lifetime. Table 16.2 displays typical values of the LCOE for plants planned to enter service in 2027. The total LCOE is the sum of (1) capital costs, (2) fixed costs for operation and maintenance costs (O&M), (3) variable costs, and (4) transmission costs.

The sum of the capital, fixed operation and maintenance, and transmission costs constitute the fixed cost, while fuel accounts for nearly all of the variable cost. Renewable energy's variable costs are negligible since there is no fuel. The number of MWh produced over the lifetime of a plant is equal to its installed capacity (in MW) multiplied by its capacity factor and by its cost recovery period (in hours). The table includes the capacity factors. Note that if the capacity factor increases, the levelized fixed costs decrease, while the levelized variable costs remain constant.

**Table 16.2** Estimated levelized cost of electricity (LCOE) in $/MWh[a]

| Plant type | Capacity factor (%) | Levelized capital cost ($/MWh) | Levelized fixed O&M ($/MWh) | Levelized variable cost ($/MWh) | Levelized trans. cost ($/MWh) | Total LCOE ($/MWh) |
|---|---|---|---|---|---|---|
| Dispatchable | | | | | | |
| Nuclear | 90 | 60.71 | 16.15 | 10.30 | 1.08 | 88.24 |
| Coal-supercritical | 85 | 52.11 | 5.71 | 23.67 | 1.12 | 82.61 |
| CC gas turbine | 87 | 9.36 | 1.68 | 27.77 | 1.14 | 39.94 |
| OC gas turbine | 10 | 53.78 | 8.37 | 45.83 | 9.89 | 117.86 |
| Renewable | | | | | | |
| Solar PV | 29 | 26.60 | 6.38 | 0.00 | 3.52 | 36.49 |
| Wind onshore | 41 | 29.90 | 7.70 | 0.00 | 2.63 | 40.23 |
| Wind offshore | 44 | 103.77 | 30.17 | 0.00 | 2.57 | 136.51 |
| Hydroelectric | 54 | 46.58 | 11.48 | 4.13 | 2.08 | 64.27 |

[a] From table 1a, U.S. Energy Information Administration | Levelized Cost and Levelized Avoided Cost of New Generation Resources AEO2020, https://www.eia.gov/outlooks/aeo/pdf/electricity_generation.pdf

The levelized cost components in Table 16.2 are based on a weighted average cost of capital financed at 6.2% and a cost recovery period of 30 years. If the cost recovery period were increased to the 60 years for which existing reactors are now being licensed, their levelized fixed costs for nuclear would be cut in half, making them cost competitive with wind and solar energy. A recent California study of offshore wind[1] and another at MIT of 1000 MW power reactors[2] both result with LCOEs of roughly 106 $/MWh, employing 30 year cost recovery period. If the fossil fueled plants were required to incorporate carbon capture to decarbonize them, estimates are that their levelized capital costs would nearly double from the numbers tabulated in Table 16.2.[3] Note that LCOE calculations do not consider the costs of backup peaking plants required to counter the erratic weather-dependent intermittency wind and solar energy. We return to the intermittency issue later in the chapter.

The economics of the two types of gas turbines differ. In either type, methane is burned to produce electricity, but a hot exhaust gas is also produced. In a combined cycle (CC) gas turbine, the turbine's exhaust gas provides heat to power a steam turbine. This leads to a higher efficiency than for an open cycle (OC) gas turbine for which the exhaust gas is simply vented to the environment. OC gas turbines,

---

[1] Schrupp, Kenneth (2024). "Newsom Orders Review of Why CA Electricity is Expensive with Renewable Mandate" thecentersquare.com.

[2] https://www.world-nuclear-news.org/Articles/AP1000-remains-attractive-option-for-U.S.-market-say

[3] Herzog, H. M. (2018). *Carbon Capture*, MIT Press.

however, can ramp up and down in power much faster than CC gas turbines. The result is that CC gas turbines are best used to provide baseload power 24/7 while OC gas turbines are employed more often as peaking units, only running a few hours per day, as indicated by the capacity factor of 10%. Calculations of LCOE in $/MWh employ the number of MWh expected to be produced over the life of the plant. Since the number of MWh produced resides in the denominator of the levelized Capital and O&M costs, the levelized costs for OC turbines (i.e. the $/MWh produced) are higher than for CC turbines, even though CC turbines are more complicated and thus are far more expensive (i.e. in $/MW) to build.

LCOE came into use as a relatively straight forward tool to calculate MWh costs when electricity was provided predominately by fossil fuel plants at vertically integrated utilities. The variable cost component of LCOE of a power plant is also called a plant's marginal cost. A utility would rank its plants' marginal costs. For each time interval, usually 1 h, the plants with the lowest marginal costs would operate, with just enough plants operating to meet the demand. The plant operating with the higher marginal cost would load follow; it would operate with constantly changing power to exactly satisfy demand. If the marginal cost were yet higher, it would operate as a peaking unit, or not at all. Two challenges have come about that have caused radical changes in the economics of electric power and decreased the efficacy of using LCOE numbers solely in estimating the cost of electricity. The first was the deregulation of electricity markets in the 1990s. The second was the rise of wind and solar energy in the first decades of the twenty-first century. We treat them in that order.

***

Traditionally, electric grids were operated by vertically integrated monopolies. Frequently in the USA, generation, transmission, and distribution were owned by a single investor-owned business, such as the Commonwealth Edison whose geographic area stretched over the Chicago metropolitan area and much of northern Illinois. In a few situations, a government entity such as the Tennessee Valley Authority, or TVA, generated and transmitted electricity to municipal or privately owned distribution utilities. In either case, there was integrated resource planning. The generating utility would project the future growth of electricity demand, evaluate the age and capabilities of its existing power plants and then determine what type and how many—if any—power plants needed to be built to meet future demand. The utility would present its plan for future power plants and transmission lines for approval by a state utility board or a federal agency charged with monitoring its operations. The board would then allow the utility to set electricity rates such that it could make a reasonable return on long-term investments. Without such guarantees, utilities would be unlikely to attract enough investors to build the needed power plants and transmission lines.

Such regulated utilities tended to emphasize reliability over affordability. Sometimes they encouraged over investment—so-called "gold plating"—because the larger the investment, the greater the regulated financial return. There were few blackouts, but many thought the rates paid by consumers were too high. In the 1990s, governments attempted to remedy the situation by deregulation, such as discussed in

Chap. 3. New rules, set by the Federal Energy Regulatory Commission or FERC,[4] led to geographical areas accounting for about two-thirds of national electricity consumption being incorporated into seven regional grids, four of them spread over multiple states. On Fig. 2.2, only the regions designated as Northwest, Southwest, and Southeast were not deregulated. Nonprofit organizations, known as Regional Transmission Organizations (RTO) or Transmission System Operators (TSO), were created to operate the seven grids under rules established by FERC. RSOs and TSOs are often referred to as balancing authorities, since it is their responsibility to balance supply and demand, that is, to ensure that they are always equal.

Within each of the seven deregulated grids, vertically integrated utilities were required to sell their power plants, hence becoming distribution utilities. The power plant owners then became known as merchant generators. The high-voltage transmission networks, while still privately owned, became regulated monopolies under the control of the nonprofit grid operators. On each grid, merchant generators compete in a wholesale market run by the balancing authority—the RTO or STO—under FERC rules. The distribution utilities—often what remained of a vertically integrated utility—purchase electricity on the wholesale market and operate lower voltage transmission networks to dispense electricity from substations to retail customers.

Nearly all grid balancing authorities oversee three wholesale markets: energy, ancillary services, and capacity. Of these, the energy market is the largest. In it, the balancing authority makes day-ahead estimates of hour-by-hour demand. Merchant generators bid how much power they can supply for each hour and at what marginal price. The marginal price is the generator's cost of providing the next MWh of electricity for a plant that is either already running or can start supplying electricity on very short notice. Averaged over time, it is comparable to the variable cost listed in Table 16.2.[5] Note that the marginal cost does not reflect the large capital cost of building carbon-free power plants, whether they be nuclear or VRE.

An hour-by-hour auction carried out the day before determines which plants will be online supplying electricity at any given hour. As illustrated in Fig. 16.1, the power plants are put in "merit order" of their marginal cost of producing power. The solid and dashed stair step lines indicate, respectively, the energy market supply with and without significant VRE—labeled renewables—operating. At any given hour, the width of each stair represents how many MW of renewable, nuclear, coal, combined cycle gas turbine (CCGT), and open cycle gas turbine (OCGT) power is available for sale, and heights of the stairs indicate the prices. The point where the supply line and the nearly vertical demand line cross sets the price paid to all the energy suppliers. It is termed the market clearing price. Note that when the production from renewables is high, shown by the solid supply line, the market clearing

---

[4] https://www.rff.org/publications/explainers/ferc-101-electricity-regulation-and-the-federal-energy-regulatory-commission/

[5] In the USA, at present, combined cycle gas turbines have a lower marginal price than coal.

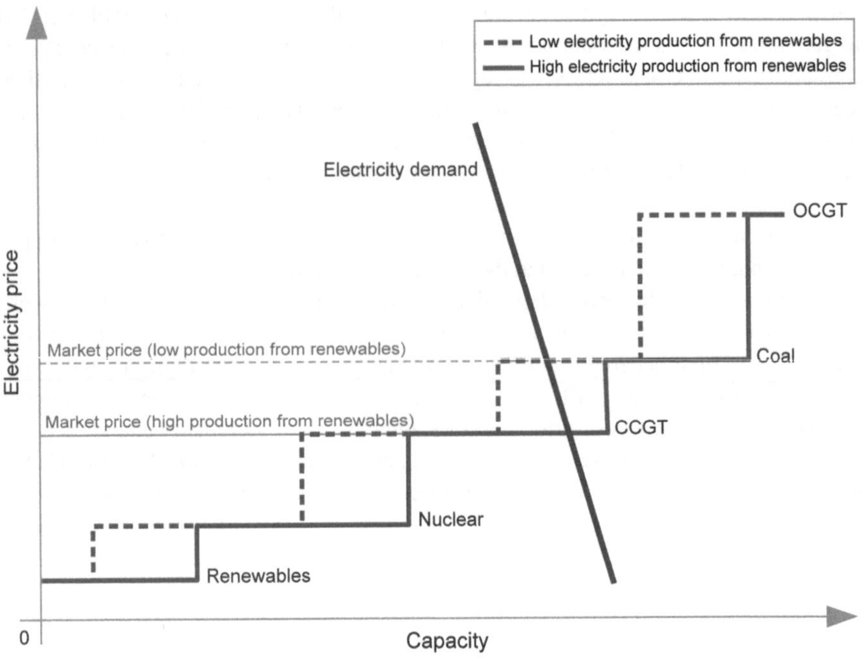


**Fig. 16.1** Order of merit graph. (Source: Nuclear Energy Agency)

price is lower than when renewable production is low, shown by the dashed supply line.

All the generators that are running receive the market price, that is, the marginal price of the most expensive plant running. Moreover, the most expensive plant running is curtailed to only generate enough energy to meet the demand. For this example, the case of high renewables production, the most expensive plant running would be CCGT. In the case of low renewables production, it would be coal. Accordingly, in these two situations, CCGT and coal, respectively, recover only their marginal costs, which are predominately for fuel. The plants with lower marginal costs also recover the difference between the market price and their marginal cost.

Hour by hour, the near-vertical demand line will shift right for higher daytime demand and left for lower nighttime demand. The stairstep supply curve tends to be more static except being moved right and left frequently by the intermittent availability of the renewable energy. Under conditions in which renewables (VRE) can supply the entire demand, the price paid for electricity is very close to zero. In some cases, it becomes negative.

In deregulated energy markets under FERC rules, the difference between market price and marginal cost is the only money available to recover capital costs and for profit. These market rules are more appropriate for fossil fuels, for much of their total cost is the cost for fuel. But they cause problems for carbon-free power plants since both renewables and nuclear have high capital costs but very small marginal costs. With increasing amounts of VRE supply becoming available, much of the time only VRE and nuclear plants are needed to supply the demand, causing nuclear plants to generate no revenue for capital cost reimbursement.

Under such rules, only when a severe weather event or other situation causes demand (and therefore the supply price) to climb markedly, do renewable plants, and to lesser extent nuclear plants, receive larger revenues for capital cost recovery. In such circumstances, public objections abound to what are dubbed windfall profits. But in fact, the plant owners rely on these funds, for under current market rules they are the principal source of money for recouping capital costs.

The dichotomy between the costs of producing electricity and the pricing of energy markets under FERC rules is made obvious by the values of LCOE listed in Table 16.2. The total LCOE measures the price that must be charged to recover the costs of building power plants and of operating them over their design lives of from three to six decades. Yet, energy market is based on marginal cost, which is primarily fuel cost. This quandary has been dubbed the "missing money problem."[6]

Wind and solar power plants as well as nuclear reactors fall into the high capital investment category. However, the subsides from federal investment and production tax credits and the renewable energy portfolio requirements are keeping the VRE boom afloat within these deregulated markets. If the US subsidies were dropped, we would see the same dearth of VRE construction that has taken place with reactors since the creation of wholesale electric energy markets in the 1990s. Likewise, it is to be expected that proposals currently being made to build modular and advanced nuclear reactors are focused on those parts of the county that retained integrated utility monopolies. The net effect of deregulation has been to make the grids less reliable and rolling blackouts or worse more likely, even while they have made electricity more affordable on average, but with much more volatile pricing.[7]

The ancillary services markets deal with very short-term challenges—from small fractions of a second to minutes—of keeping the grid stable. Stability is provided mostly by the presence on the grid of the synchronous inertia of the large spinning masses of turbine generators, since they resist and slow changes in frequency and voltage. As the fraction of generation provided by VRE increases, however, the proportion of turbine generators deceases, causing the grid to tend toward instability. Thus, the need for ancillary services increases, for they are designed to provide enough supplementary power to quickly compensate for a sudden power plant or transmission line failures until slower responding generators can supply the needed power. The markets contract for fast-acting backup services such as turbine generators operating at partial power to provide spinning reserve. Battery storage and OC gas turbine or piston powered generators that can quickly ramp to full power are also contracted for ancillary services.

---

[6] https://www.nrel.gov/docs/fy15osti/64324.pdf

[7] https://kleinmanenergy.upenn.edu/podcast/grid-forward-debate-has-electricity-deregulation-led-to-better-community-outcomes/

While the ancillary services market deals with minute-to-minute phenomena, the capacity markets deal with multiyear durations. They provide payment to merchant generators who guarantee to have their plants ready of operate if needed, but only for three or in some cases only 1 year into the future.[8] Such capacity markets encourage construction of OC gas peaking units, which are faster and cheaper to build than other power plants. In principle, capacity markets also should assure long-term grid reliability. They should compensate for the energy markets' lack of mechanisms for securing adequate revenue to recoup the capital costs of the baseload power plants that will be needed to provide electricity in the decades to come. Capacity markets' short time horizons, however, fail to ensure the long-term availability of adequate generating capacity. As presently structured, they don't offer a solution to the missing money problem. But a solution must be found if carbon-free power plants are to be built, for virtually without exception they are capital intensive, while having little or no marginal costs.

Governmental agencies in a number of nations are alleviating the missing money problem by utilizing various forms of the contracts of difference discussed in Chap. 3. In the USA, there are no such federally operated programs. However, high-tech companies are setting up analogous arrangements to secure long-term baseload supplies of electricity to power large data banks needed for artificial intelligence projects. Power Purchase Agreements (PPAs) are financing the restarting of the Palisades, Three Mile Island 1, and possibly other shuttered reactors. They may also be used to buy power from existing reactors and to finance small modular reactor construction. In its simplest form, a PPA is a long-term contract between a power plant owner and a purchasing company in which electricity is sold at a set price over a fixed time period, typically 20 years. On deregulated grids, when the wholesale market price is higher than the set price, the plant owner pays the purchasing company the difference. Conversely, when the market price is lower than the set price, the purchaser pays the plant owner the difference. The net result is that electricity is sold at the set price protecting both buyer and seller from the volatility of wholesale electricity markets.

***

Before the integration of VRE onto electric grids, using LCOE numbers—such as those in Table 16.2—gave fairly accurate costs of individual power plants. They were direct costs, for the most part independent of the other plants on the grid. It was fairly straight-forward to compare costs and decide which would be most economical to operate at any given time. The LCOE, however, does not account for all the cost of adding a new generator to the grid. It accounts for only the plant-level costs directly attributable to the individual plant. As Fig. 16.2 indicates, there are also grid-level as well as social and environmental costs. Grid-level costs are incurred when any power plant is added. Often called integration costs, they are

---

[8] https://www.gao.gov/assets/gao-18-131.pdf

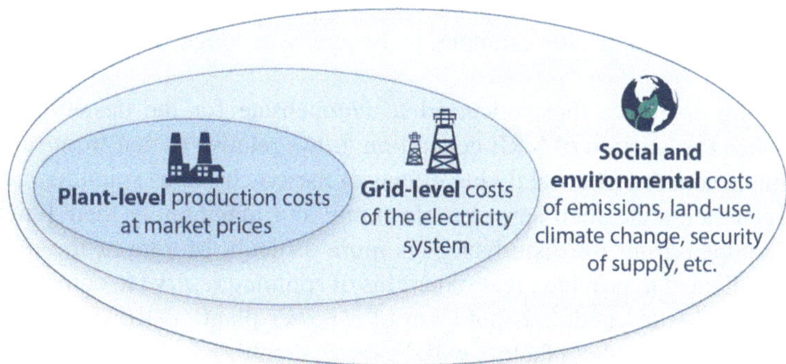


**Fig. 16.2** Major cost categories in electricity systems. (Source: Nuclear Energy Agency)

much larger for VRE than for dispatchable power plants.[9] They fall roughly into three groupings: profile, balancing and grid costs.[10]

Profile costs are caused by VRE supply profiles differing greatly from demand profiles, for VRE generation is often erratic, diminished, or not available at all. The difference in profiles is exemplified by comparing the VRE supply profiles in Fig. 5.2 with the typical demand profile of Fig. 5.3. As detailed in the foregoing section, energy markets are based on the balancing authority making day-ahead estimates of demand on an hourly basis, and on merchant generators bidding on how much of the demand they can supply during each hour. Unlike dispatchable energy sources, VRE supply will vary from hour to hour with the weather. Moreover, most of the time VRE will either produce too much power, requiring curtailment, or too little, requiring dispatchable backup energy from nuclear, fossil fueled, or hydroelectric generators to make up the difference. To supply the residual demand as VRE grid penetration increases, dispatchable power plants need to make frequent large changes in power, ramping up or down, and often operating at small fractions of full power for extended durations. Such operations lead to dispatchable power plants having reduced capacity factors, increased maintenance, wear and breakdowns, thus resulting in large increases in their LCOE.

Energy storage is often considered as an alternative to backup power plants, and indeed battery banks are used increasingly to fill in momentary gaps in supply and to smooth the transition from day to night on grids heavily dependent on solar energy. At full power, however, the lithium-ion battery banks discharge in 4 h.

---

[9] Hirth, L. et al. (2015). "Integration costs revisited—An economic framework for wind and solar variability," *Renewable Energy* 74, 925.

[10] The Costs of Decarbonization: System Costs with High Shares of Nuclear and Renewables, © OECD 2019 NEA No. 7299.

Furthermore, no known method of storage even comes close to being economically viable for supplying large quantities of electricity over longer durations to compensate for the sun at night, for example, or for a day or longer for overcast skies or calm winds.

Balancing costs are those incurred to compensate for the destabilizing grid effects when the fraction of VRE generation grows relative to that from stabilizing dispatchable sources. The stabilizing effects of the synchronous spinning masses of turbine-generators are then spread more thinly and maintaining their function of rapidly compensating for disturbances is more difficult. Moreover, the remaining turbines will need to provide greater margins of spinning reserve to compensate, for example, for the unscheduled shutdown of a power plant—VRE or dispatchable. Consequently, turbine load factors will decrease, resulting in an increased LCOE.

To compensate for grid loss of stability, the strategy sometimes considered is to add synchronous condensers—also called synchronous compensators—to the grid. Each is the equivalent of a large generator, synchronized to 60 Hz, but operated as an idling motor. A heavy flywheel may be attached to the generator shaft to increase its inertia. These are large expensive pieces of equipment that would add substantially to grid costs.

Grid operators also incur costs in connecting power plants to existing high-voltage transmission lines. Most often nuclear and fossil fueled plants are located much closer to the population centers where electricity use is concentrated than are VRE farms. In the USA, VRE from plants already built is being curtailed by the unfilled need for thousands of miles of high-voltage transmission lines to carry VRE electricity from remote locations to population centers. Moreover, once built, such lines must be capable of carrying the maximum power produced by VRE plants. Intermittency, however, will mean that much of the time they will carry only a fraction of their capacity, or none at all. Thus, relative to transmission lines carrying dispatchable electricity, lines transmitting power from VRE farms will be underutilized; they will have smaller capacity factors, driving up capital cost per megawatt mile of electricity transmitted. For offshore wind, the connection to the land-based grid is particularly expensive, for it involves laying miles of transmission lines on the ocean bed which will be used solely to transmit wind energy.

Outside of plant- and grid-level costs, Fig. 16.2 shows a larger area, labeled Social and Environmental Costs to account for expenses not included in the other two categories. They are costs not borne by the plant owners or grid balancing authorities, but by the public. Foremost among environmental costs are those stemming from carbon dioxide emissions from fossil fuel plants. In the absence of adequate supplies of hydroelectric or nuclear energy, even as the $CO_2$ from coal combustion is reduced, a growing number of fast-start gas turbine generators will be needed, operating in start-stop mode to accommodate VRE intermittency. Operated in this manner, they emit nearly double the $CO_2$ per MWh than they would be running at constant power.[11] The poor combustion from continually ramping gas-powered plants also produces excessive amounts of carbon monoxide, nitrous

---

[11] http://euanmearns.com/co2-emissions-variations-in-ccgts-used-to-balance-wind-in-ireland/

**Table 16.3**  Direct and system levelized costs of electricity (LCOE) in \$/MWh[a]

| Plant | Direct | System | Total |
| --- | --- | --- | --- |
| Nuclear | 82 | 122 | 204 |
| Wind | 40 | 291 | 331 |
| Solar | 36 | 413 | 449 |

[a] https://advisoranalyst.com/wp-content/uploads/2023/05/bofa-the-ric-report-the-nuclear-necessity-20230509.pdf

oxides, sulfur dioxide, methane, volatile organic compounds, and particulate matter, some of which are both ground-level pollutants and greenhouse gases. Other environmental degradation may take the form of water pollution, endangered species destruction, and unsightly energy sprawl.

The foregoing discussion illuminates the many grid and environmental costs that are not factored into LCOE tabulated in Table 16.2. They are much larger for VRE than for plants producing dispatchable power for three reasons. First is the cost of building plants to provide backup power. These must be capable of ramping up and down in power rapidly. If fossil-fueled, they require carbon capture and sequestration to prevent greenhouse gas emissions. Second is the cost to compensate for the loss of grid stability and the inability to counter disruptions that accompany VRE penetration. Third are the costs of the thousands of miles of high-voltage transmission lines needed to carry electricity from remote VRE locations to population centers. While storage technology may someday compensate for some of these costs, then the storage costs must be attributed to VRE.

Much is said about the high capital costs of nuclear plants making them uncompetitive with VRE. Indeed, as shown in Table 16.2, if only direct (i.e., plant level) LCOEs are compared, the total nuclear LCOE is more expensive those for wind and solar energy. However, if plant- and grid-level costs are combined to give system LCOE values, then results of a recent Bank of America Global Research study, shown in Table 16.3, indicate that nuclear power is the less expensive than either wind or solar energy.

***

The cost of an electric grid cannot be determined simply by adding the levelized costs—the LCOE—of each of its constituent generators, transmission lines, substations, and other components. They interact in multiple ways that must be taken into account to assure that the grid is stable and balanced at all times, even while grid capacity is increased rapidly to accommodate growing electricity demand, and while formidable efforts are undertaken to eliminate fossil fuel combustion. Attracting private investment in the amounts needed to build nuclear plants is made more difficult by their cost and longevity. We now know that they have operational lives that can be expected to last from 60 to 80 years, and possibly longer.[12] In sharp contrast, wind and solar farms have design lives of only 20–30 years. Hence, nuclear

---

[12] https://www.iaea.org/newscenter/news/going-long-term-us-nuclear-power-plants-could-extend-operating-life-to-80-years

investors need to consider the uncertainties of much longer time horizons than those investing in intermittent renewable energy.

The wholesale electric markets under which much of the USA currently operates provide little if any incentive for the large long-term capital investments needed for thousand MW sized reactor construction, exacerbating the missing money problem. Even in the parts of the country where investor-owned public utilities still exist, the capital requirements and long cost recovery periods are problematic. It should not be surprising that much of the world doesn't rely primarily on private investment to build large multiunit nuclear power plants. Nearly all such plants being constructed are under the auspices of state-owned or controlled utilities that have the power to set electric rates and thus to assume large shares of the risk of large long-term investments.[13]

In the USA, of power sources that require huge capital investments and have very long lives, one stands out: the large dams that provide hydroelectric power. Of these, the 395 MW power station at the Hells Canyon dam is the largest that is privately owned.[14] All larger hydroelectric plants are operated by the Bureau of Reclamation, the Bonneville Power Authority, the Army Core of Engineers, the Tennessee Valley Authority or other federal agencies, or by state agencies such as the New York Power Authority, which operates the Niagara Falls hydroelectric plants. Such governmental entities provide the financial resources and assume the risks of large energy projects with very long payback periods. For example, the Hoover Dam's construction was financed by a 50-year mortgage from the US Treasury to the Federal Bureau of Reclamations, which has responsibility for the dam and the operation of its hydroelectric plant. There is also precedent for a federal agency financing reactor construction. The Tennessee Valley Authority has assumed responsibility for building nuclear reactors as well as fossil fueled and hydroelectric plants.

If the USA is to become carbon neutral, fulfilling its commitments to stem climate change, tens of billions of dollars will be required to build large nuclear power plant complexes with capital reimbursement periods of well over half a century. Even with reformed FERC rules for electricity markets, raising adequate capital primarily from private investors to finance such projects seems problematic. Government backed incentives, such as those in the UK and Australia outlined in Chap. 3, may help as may power purchase agreements made between plant owners and purchaser of power in the USA. Nevertheless, the challenge of curbing climate change raises the question of whether in some form, the US government needs to step in.

---

[13] https://world-nuclear.org/information-library/economic-aspects/financing-nuclear-energy

[14] https://oemr.idaho.gov/sources/re/hydropower/

Electric grids as much as the interstate highway system, airports, water treatment, waste disposal, and more are basic infrastructure upon which society depends. Yet, while the financing of other infrastructure is backed by the taxing powers of governmental agencies at various levels, electric grids for the most part rely on private investment. Based on integrated resource planning, a new federal agency, or those presently operating large hydroelectric plants, could be provided with the authority to finance large decarbonization projects. Such an agency or agencies would take on responsibilities similar to those of the government-backed organizations responsible for building hydroelectric plants and most of the world's large nuclear power plants.

# Chapter 17
# Forecasting the Future

**Keywords** Grid comparisons · Renewables dominant grids · Nuclear dominant grids · Systems analyses results · Cost comparisons · Land requirements · Grid stability · Blackout risks · Dark doldrums

All indications are that demand for electricity will increase rapidly in the coming decades. The need for new generation capacity and for transmission lines to deliver electricity to where it is consumed will escalate even faster. Concomitantly, fossil fuel plants must be shut down and replaced with decarbonized power plants and energy for transportation and other economic sectors converted to electricity to meet the goal of carbon neutrality by 2050. The effects of the emergence of artificial intelligence on electricity demand are unclear. The immediate impact of its energy-consuming data centers is to drive demand upward, but its application to electric grids and other industries may mollify the increases, possibly reducing the rate of demand growth.

Energy conservation, biofuels, smarter electric grids, and more certainly play significant roles in reducing greenhouse gas emissions. But in the end, renewable and nuclear energy together will carry the lion's share of the load. In the limited geographical areas blessed with necessary resources, hydroelectric or geothermal power may provide a large share of electricity supply. But for most of the world, wind and solar energy will be the dominant forms of renewable energy. Thus, the question is should these forms of variable renewable energy (VRE) or nuclear energy play the leading role? To pull together what we have learned from the foregoing pages to answer this question, we devote this final chapter to examining the characteristics and implications of future grids dependent principally on VRE and on nuclear energy.

***

Wind turbine and solar panel costs have plummeted since the turn of the twenty-first century, generating enthusiasm that future electric grids can be based mainly or entirely on renewable energy. Concurrently, the tools of systems analysis have greatly advanced, allowing estimates of the characteristics of future grids to be



E. E. Lewis, *Renewables or Nuclear*, https://doi.org/10.1007/978-3-032-08074-5_17

refined. The analyses use decades of data of hourly weather, electricity demand, and economic variables. These increase the confidence with which forecasters may hold the resulting predictions. With some exceptions, such studies focus on electric grids becoming 100% renewable, with nuclear barely mentioned, if at all. Most studies also assume that enough land and transmission line capacity will be available to achieve the renewables goals. We briefly review available studies before considering the conclusions to be drawn.

A comprehensive study comparing different levels of wind and solar energy was carried out at MIT under the auspices of the OECD.[1] It involved a 1-year simulation of a 67 gigawatt (GW) grid that included many power plants. French demand curves and weather data on an hour-by-hour basis provided necessary input. The study placed a cap on emissions of 50 kg of carbon dioxide per megawatt hour of electricity ($CO_2$/MWh), which is about one-eighth of current US emission levels. The five scenarios considered were for 0, 10, 30, 50, and 75% VRE. The amount of hydroelectric generation was held constant, while the proportions of fossil fuels, nuclear, wind, and solar energy were optimized to minimize cost in each of the five cases.

Results are summarized in Figs. 17.1 and 17.2. Figure 17.1 shows the distribution between the sources of energy. The base case has no VRE, while nuclear, combined cycle gas turbine (CCGT), and hydroelectric power provide 80, 11, and 9% of the power, respectively. While nuclear energy dominates the base case, it decreases in energy output as the percentage of VRE increases, and nuclear is absent from the 75% VRE case. A 100% VRE case is impossible because considerable dispatchable capacity, largely in the form of open cycle gas turbines (OCGT), must be present to compensate for VRE's intermittency, and the study did not allow unrealistic expansion of hydroelectric generation. Even though coal has the lowest cost of energy, no results appear for it because the 50 kg of $CO_2$/MWh cap could not be met if coal was included in any of the scenarios.

Figure 17.2 shows the total installed capacity increasing with the percentage of VRE. This is expected since the capacity factors for wind and solar generation are quite low compared to the other power sources. Yet, the climb is striking, from 98 GW in the base case to 220 GW at 50% VRE, and 325 GW at 75% VRE, more than three times the base case. Perhaps, the study's more important result is not illustrated by the figures. The total cost increases as the fraction of VRE increases by 5, 21, 42, and 98% for 10, 30, 50, and 75% VRE generation. Thus, costs nearly double in going from 0 to 75% VRE generation.

The study revealed several additional phenomena. As VRE percentage of capacity increases, gas and nuclear plant as well as system capacity factors decrease. Dispatchable power ramping speed requirements increase as VRE capacity increases. With 50% VRE, nuclear units must ramp up and down by 30–35% of capacity in 1 h. Nuclear is excluded from 75% VRE because the ramping requirements are too stringent for present-day power reactors. Curtailment of VRE starts

---

[1]THE COSTS OF DECARBONISATION: SYSTEM COSTS WITH HIGH SHARES OF NUCLEAR AND RENEWABLES, NEA No. 7299, OECD 2019.

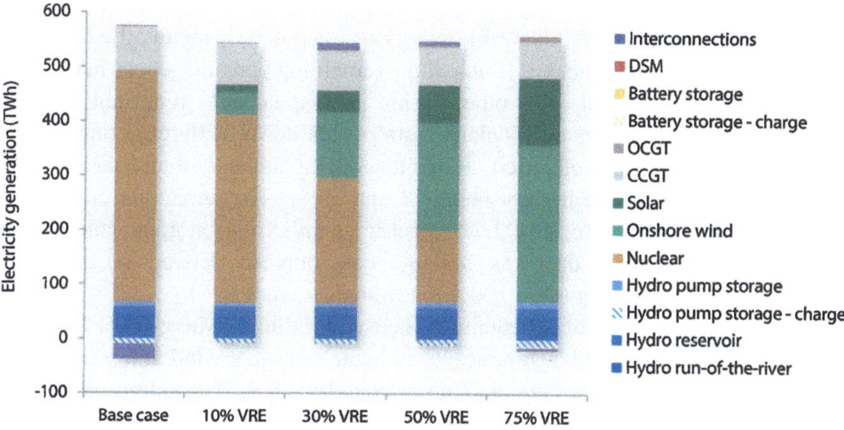


**Fig. 17.1** Comparison of varying mixes of generating shares. (Source: Nuclear Energy Agency)

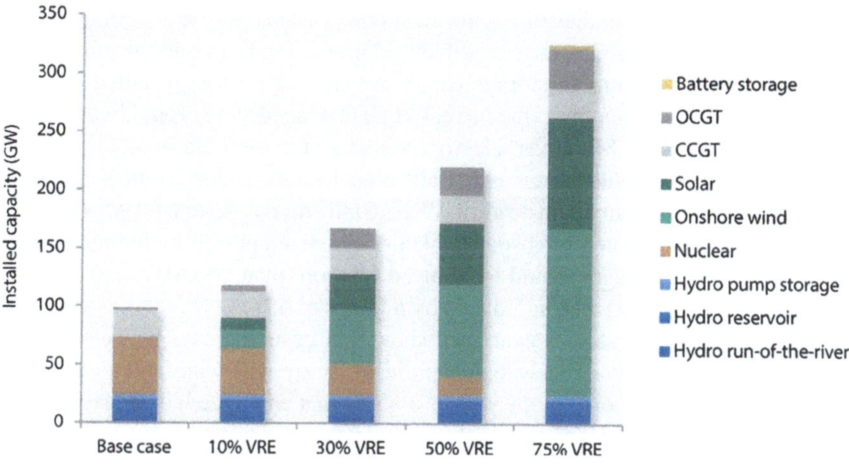


**Fig. 17.2** Comparison of varying mixes of generating capacity. (Source: Nuclear Energy Agency)

when it consists of 30% of capacity. At 50% VRE capacity, 10% of it is curtailed, and at 75% VRE, 18% of it must be curtailed.

The foregoing study didn't allow for future advances in technology, such as large advances in energy storage capacity or improved reactor ramp rates. Nor did it allow for demand management or related smart grid techniques. The systems' analyses that we next consider are more futuristic, allowing for technological advances incorporated in smart grids, energy storage, and more. Nevertheless, examinations of deep VRE penetration yield similar results.

Focused on the USA, a study of Renewable Energy Futures found that a future grid optimized to have 80% renewable energy resulted in only about 50% of it coming from wind and solar energy.[2] Much of the remaining percentage was made up of hydroelectric, geothermal, and other forms of dispatchable renewable energy. Similarly, an examination of 15 plans from a half dozen different countries for reaching 100% renewable electricity found that only 2 had more than 65% wind and solar generation, with the highest being 73 and 69%.[3] The remaining energy was obtained from combinations of hydroelectric, geothermal, storage, and energy imports. In studies where data was available, cost increases between 41 and 104% were incurred in achieving these 100% renewables scenarios.

An examination covering 42 countries focused on studies where 100% VRE was assumed. However, net 100% renewables indicate only that wind and solar energy produce an amount of energy equal to the demand for a year. The definition does not require that electricity be produced when needed. Consequently, the investigators concluded that with annual VRE generation equal to annual demand, intermittency would cause VRE to produce excess power over significant time spans. More important, VRE supply would fall short of demand for time periods totaling nearly 800 h per year.[4]

Three independent studies of California's power needs growing from 50 to 100 GW by midcentury were in consensus that "solar and wind can't do the job alone."[5] Even with dramatic gains in storage capability, at 100% VRE, installed capacity would reach 500 GW, ten times the increased 50 GW needed and roughly that of the entire USA at present. Moreover, electricity cost would increase by 46%, and it is unlikely that there would be enough California land available to meet the 6250 square mile requirement. If instead of VRE, California's present 42 MW of gas-fired plants were displaced by dispatchable decarbonized power, including nuclear, the need for VRE capacity would be reduced by more than 400 MW, and the need for transmission line expansion reduced by a factor of 4.[6]

Taken together, the studies examined show that costs increase between 41 and 104% in achieving 100% renewables, and in no study was wind and solar power stated to exceed 75%. Some of the studies assume that enough wind and solar farms could be built to supply 100% of *net* electricity demand. As stated above, net means only that averaged over a year's time the total electricity supply will equal the total demand. That isn't the primary challenge, however, for at any point in time the weather-driven intermittency of wind and solar energy will cause their supply to be either greater than or less than demand. Hence, without a massive increase in

---

[2] https://www.nrel.gov/docs/fy13osti/52409-ES.pdf

[3] Deason, W. (2018). "Comparison of 100% renewable energy system scenarios with a focus on flexibility and cost," *Renewable and Sustainable Energy Reviews*, 82, 3168.

[4] Tong, D., et al. (2021) "Geophysical constraints on the reliability of solar and wind power worldwide." *Nat Commun* **12**, 6146.

[5] Long, Jane C.S., et al. (2021). "Clean Firm Power is the Key to California's Carbon-Free Energy Future." *Issues in Science and Technology,* National Academy Press.

[6] Ibid.

long-duration energy storage—something that no known storage method can deliver—substantial quantities of dispatchable backup energy will be necessary to redistribute power to when it is needed. Such backup is expensive, and where hydro-electric or geothermal power is not abundant, it must come from nuclear or other nonrenewable sources.

How do the results of these studies relate to the levelized cost of energy (LCOE) examined in Chap. 16? Recall that LCOE is just the price that must be charged for electricity in dollars per megawatt hour ($/MWh) to recover the costs of building and operating a power plant over its operational life. In Table 16.3, the distinction is made between the direct and system cost components of LCOE. The direct costs are those incurred in building and operating a plant. The system costs are those that stem from integrating that plant into an existing electric grid. Total costs are the sum of the two. The table indicates that nuclear plants have higher direct cost, but wind and solar system costs are the higher.[7]

Much of the causes for the high VRE system costs are for the backup power and longer-distance transmission lines needed to integrate wind or solar energy onto a grid. The result is a lower total LCOE for nuclear than for VRE. In effect, the consumer would pay less for power produced by a reactor than for power produced by VRE that includes the costs of its necessary backup system and longer transmission lines. The foregoing studies' conclusions that electricity costs increase with VRE grid penetration are consistent with a recent survey that correlated cost in $/MWh with the percentage of wind and solar energy. Figure 17.3 shows solar and wind power generation costs across nearly 70 countries correlating with higher average

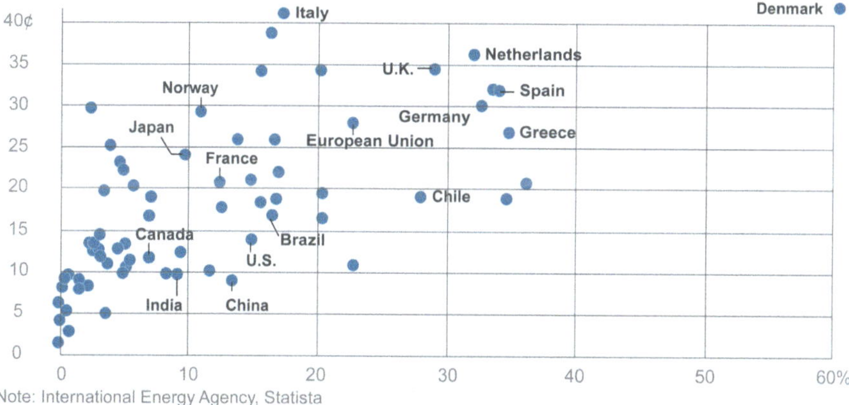


**Fig. 17.3** Average electricity price per kwh, industry and household vs percent solar and wind in electricity. (Source: Wall Street Journal)

---

[7] https://advisoranalyst.com/wp-content/uploads/2023/05/bofa-the-ric-report-the-nuclear-necessity-20230509.pdf

household and industry energy prices. For every 10% increase in solar and wind share, the electricity cost increases by more than 5 cents a kilowatt-hour.[8]

***

The foregoing studies reveal the additional costs incurred as the percentage of VRE increases on electric grids. With large grid penetrations of VRE, other considerations related to risk and reliability come to the forefront. Freezes, heat waves, hurricanes, and other severe weather episodes cause wind turbines to shut down and solar panels to be robbed of most of their power. Thus far, in encountering such weather events, fossil fueled plants have been brought online soon enough to escape dire consequences. But as shuttered fossil fueled plants are dismantled and no longer available for emergencies, the vulnerability of grids heavily dependent on variable renewable energy grows.

To grasp the uncertainties, a look at the situation in Denmark may be helpful, for Fig. 17.3 shows it to be the leader in striving to reach 100% renewable—predominantly wind—energy. Denmark's officials admit "There will be lots of complicating factors in an electricity system based 100% on renewable energy. Ensuring balance and stability in the future power system will be a huge task."[9] The article continues, "how much water will there be in the Nordic hydroelectric power stations? How windy will it be? And how long and cold will the winter be? We are sailing closer to the wind in our energy systems in Europe."[10] Worry is expressed over having fewer thermal power stations to provide grid stability and that are easy to ramp up and down. The article also stressed the need for electric consumption to become more flexible,[11] that is, to bend the demand profile to fit more closely to that of VRE supply.

As VRE penetration grows to 50% and beyond, the two fundamental requirements that electric grids must process are challenged. As detailed in Chap. 2, they are stability and balance. The dangers of instability materialized on the Iberian Peninsula on April 28, 2025.[12] Before dawn, nearly all the electricity came from a nuclear and hydroelectric plants, with the synchronous inertia of their steam and water turbines maintaining a high degree of grid stability. However, as the sun rose, the reactor power was cut back, and hydroelectric plants were shut down almost completely to accommodate the copious amounts of solar energy coming onto the grid. As the noon hour approached, the 32 GW being produced far exceeded the 25 GW demand. The excess was exported to other grids and used to fill pump storage reservoirs. Concomitantly, telltale signs of instability appeared in the form of grid frequency wavering about the synchronous 50 Hz, accompanied by oscillatory deviations in voltage.

The grid was in its most vulnerable state. VRE, predominately solar energy, accounted for more than 70% of the electricity production. Synchronous inertia had

---

[8] Lomborg, B. (2025). "Green Electricity Costs a Bundle" Wall Street Journal, January 15.

[9] https://en.energinet.dk/media/ht2lfiei/energinet-årsmagasin-2023-en-online.pdf

[10] Ibid.

[11] Ibid.

[12] https://inesctec.substack.com/p/the-iberian-peninsula-power-grid

plummeted, now supplied by four reactors operating at fractions of full power and from hydroelectric and gas-fired peaking plants. A fifth reactor had been shut down much earlier for refueling, while two others were offline because their marginal cost of producing electricity was higher than the slightly negative cost for wind or solar power at that time.

For reasons still being debated, efforts to quell the voltage oscillations resulted in an excessive voltage increase that at 12:23 caused three plants in an area dominated by solar electricity production to shutdown in rapid secession.[13] The resulting supply–load imbalance caused the grid frequency to decrease to where all other plants shut down automatically; they shed load to prevent irreparable damage from occurring. The cascading shutdowns caused the grid to collapse into blackout, and it all happened within 1 min! Shortly thereafter, a hydroelectric plant, which has black start capability, initiated the process of bringing geographic areas back on line one-by-one until power was restored nationwide. Within 10 h, 40% of demand was met and 99% within 16 h.

The April 28 blackout made plane the risks of instability incurred from high grid penetration of wind and solar energy. Imbalances from VRE supply falling below demand for extended time periods may result if far worse blackouts. Dark doldrums—called dunkelflaute in German—are periods of overcast skies and calm winds during which VRE generation plummets to tiny fractions of its installed capacity. Dunkelflaute that are localized or that last less than a day or so are most likely manageable. Those that are widespread and long lasting, however, will be disastrous for grids dependent principally on wind and solar energy. Severe dunkelflaute are rare but inevitable. Consider the following two examples.

In January 2017, dark doldrums covered nearly all of Germany for 9 days.[14] Wind and solar installations combined produced only about 5% of their installed capacity during the dunkelflaute. At times, dispatchable power was required to supply more than 90% of electricity demand. Fortunately, the half dozen nuclear plants that were still operating plus gas and coal plants were able to fill the power deficit. With the nuclear plants now shut down, and a goal of reaching 80% renewables, however, the consequences of future dark doldrums are likely to be acute.

In 2021, dark doldrums lasted 11 days in Britain.[15] At that time, wind turbines generated roughly half of its electricity, many of them offshore on the North Sea. During the doldrums, wind generation fell to as little as 3% of what it had been. Fortunately, enough gas-fired plant capacity still existed to make up most of the wind deficit, with coal and biomass plants ramped up to near full power, and nuclear plants contributing as well.[16] In the future, dark doldrums will no doubt become a more serious threat to grid reliability as VRE grid penetration grows and fossil

---

[13] https://gridstrategiesllc.com/project/a-review-of-reports-on-spanish-blackout-causes-and-solutions/

[14] https://www.next-kraftwerke.com/knowledge/what-is-dunkelflaute

[15] https://reports.electricinsights.co.uk/q1-2021/when-the-wind-goes-gas-fills-in-the-gap/

[16] Ibid.

fueled plants are retired and then dismantled beyond the point where they can be restarted to deal with weather emergencies.

Widespread long-lasting dark doldrums are the Achilles heel of electric grids heavily dependent on wind and solar energy, for they may lead to widespread long-lasting blackouts with disastrous consequences. Unlike the Spanish blackout where stability could be restored and recovery initiated shortly after its occurrence, a blackout caused by dark doldrums would last as long as the doldrums, which could be for a week or more. Foreseeable advances in energy storage, continent spanning transmission lines, load management, and other smart grid technologies are unlikely to contribute sufficiently to fend off wind and solar droughts lasting for lengthy time periods. Building power plants or storage facilities for use only during such rare events would be economic nonstarters. Thus, along with affordability and day-to-day reliability, dark doldrums should be a central concern in policymakers' decisions as to whether renewable or nuclear energy should play the larger role in the fight to curb climate change.

***

In principle, nuclear plants could provide all the power required for electric grids. Their spinning turbines provide grid stability, and as has been demonstrated in France, reactors can load follow to meet time-varying demand for electricity, such as those shown in Fig. 2.7, for all seasons. To deal with contingences such as a power plant failure, several nuclear plants can operate at slightly less than full power, thus creating sufficient spinning reserve to quickly compensate for the lost power. To deal with sudden demand spikes, peaking plants may also be needed. These could be based on adapting naval propulsion reactor technology to electricity generation. Alternatively, several hours of thermal storage could be paired with modular reactors, such as in the Natrium reactor system, described in Chap. 14. In addition, if available, hydroelectric or other nonnuclear sources of carbon-free dispatchable backup power—such as open cycle gas turbines fitted with carbon capture and sequestration—could augment the nuclear generators.

Nuclear power plants have the added advantages of doing quite well in foul weather. During the 2021 Texas Freeze, three reactors functioned flawlessly, while a fourth was down for only a few days, far less time than most fossil fueled and renewable energy plants were out of service.[17] During the 2019 polar vortex, Minnesota's three reactors provided power continuously, while wind turbines froze.[18] When Hurricane Florence swept through Georgia, the Carolinas, and Virginia in 2018, wind speeds of more than 60 miles per hour forced wind turbines to shut down, and overcast skies caused solar panel output to drop. In contrast, only one of nine nuclear power plants in hurricane's path was taken offline during the

---

[17] https://freopp.org/oppblog/crisis-averted-texas-february-freeze/

[18] https://energynews.us/2019/02/27/wind-turbine-shutdowns-during-polar-vortex-stoke-midwest-debate/

storm.[19] Likewise, throughout California's record-breaking 2020 heat wave, the state's two reactors functioned at full power, while other generation—both fossil fueled and renewable—faltered, resulting in rolling blackouts.[20]

In reality, no grid is likely to approach 100% nuclear power for the simple reason that France, Sweden, Canada, the USA, and other nations planning to expand nuclear generation to decarbonize existing grids already rely on varying amounts of hydroelectric, geothermal, wind, and solar energy. Inevitably, the load-following and peaking generation required of nuclear plants are made more challenging by the presence of significant variable renewable energy generated at wind and solar farms. Instead of following the smoother variations in total demand, the nuclear plants then must follow the jagged residual—also called net—demand curves, such as illustrated in Chap. 5. Likewise, with increased VRE grid penetration, nuclear turbine generators will be called on to provide larger fractions of grid stability and spinning reserve.

Public fear of radiation, particularly of nuclear accidents, adds considerably to the costs of nuclear energy, even though, as Fig. 13.3 illustrates, the historical record shows nuclear power to be far safer than fossil fuels for generating electricity. Aside from the Chernobyl disaster, with its antiquated Russian reactor design, lax regulation, and negligent staff, no radiation fatalities have resulted from nuclear power reactor accidents. Even when the record-breaking Fukushima tsunami caused 1960s' design reactor cores to melt, no member of the public was exposed to dangerous levels of radiation.[21] However, hasty evacuations resulting from exaggerated radiation fears caused well over a thousand deaths.[22]

Reactors constructed in more recent decades include much improved safety systems, and staff qualifications have become more stringent. Nonetheless, licensing has not kept pace with these advancements, with the dubious linear no threshold hypotheses and the resulting ALARA methodology discussed in Chap. 13 still being employed. These have contributed to increased capital costs, construction delays, and frivolous lawsuits. However, efforts are in progress to modernize licensing procedures to reflect the increased safety of twenty-first-century reactor design and operation. Such modernization should ameliorate the costs of nuclear power.

The large capital costs of thousand MW, that is of GW-sized nuclear plants, aggravated in the USA and other western countries by schedule slip and cost overruns, have been a continuing challenge. However, there is reason to conclude that costs and construction times can be greatly reduced.[23] Moreover, it should be noted that schedule slip and cost overruns also impact large renewables projects. In

---

[19] https://cna.ca/2018/10/29/hurricane-florence-no-threat-to-nuclear-power-plants/

[20] https://www.utilitydive.com/news/california-releases-final-root-cause-analysis-of-august-rolling-blackouts/593436/

[21] https://world-nuclear.org/information-library/safety-and-security/safety-of-plants/appendices/fukushima-radiation-exposure.aspx

[22] Ibid.

[23] https://www.oecd-nea.org/jcms/pl_30653/unlocking-reductions-in-the-construction-costs-of-nuclear

particular, two large wind power projects, Ocean 1 and 2 totaling more than 2 GW capacity, were to be located off the New Jersey coast. They have been canceled due to supply chain problems and increased interest rates.[24] In addition, if a substantial fraction of power production is to come from VRE, additional costs will be incurred from the thousands of miles of high-voltage transmission lines that must be constructed to carry the power from the remote locations of wind and solar farms to the population centers where it is to be consumed. Even now, the transmission line shortage is dire, preventing completed wind and solar farms from being connect to electric grids, and line construction is notorious for cost overruns and schedule slip.

Much of the financial attraction of smaller modular reactors compared to the GW-sized plants currently in operation stems from two characteristics detailed in Chap. 14.[25] First, they can be ordered one module at a time as needed. Hence, the capital outlays for construction are for the most part incremental, incurred only as each module is ordered. Second, since modular reactors are smaller, much of the construction can be completed off site at factories where more stringent quality control and the efficiencies of mass production can be practiced. The result should be lower construction costs, shorter times between the start and completion of construction, and hence reduced financing cost.

Demand for new power plant capacity is forecast to more than double by mid-century as fossil fueled plants are retired and much of the transportation, process heat, and other sectors of the economy are electrified to meet the goal of carbon neutrality.[26] Whether future nuclear expansion will consist principally of many modular reactors of 300 MW or less, or a smaller number of GW-sized plants will depend on the tradeoffs between the economies of modular factory construction and the economies of scale. Moreover, since GW-sized reactors, such as the Westinghouse AP1000, are already operating, construction should not be delayed by the time needed to complete pilot plant testing of advanced design modular reactors before they can be produced in large numbers. Even with the success of small modular reactors, GW-sized nuclear power plants will likely be needed to meet the goal of electrifying the economy rapidly to achieve carbon neutrality.

---

[24] https://us.orsted.com/news-archive/2023/10/orsted-ceases-development-of-ocean-wind-1-and-ocean-wind-2

[25] https://nap.nationalacademies.org/catalog/26630/laying-the-foundation-for-new-and-advanced-nuclear-reactors-in-the-united-states

[26] Larson, E. et al.,(2021). *Net-Zero America: Potential Pathways, Infrastructure, and Impacts*, Princeton UP.

# Abbreviations

| | |
|---|---|
| AC | Alternating current |
| AI | Artificial intelligence |
| ALARA | As Low as Reasonably Achievable |
| BEIR | Biological Effects of Ionizing Radiation |
| BWR | Boiling water reactor |
| CAISO | California Independent System Operator |
| CC | Combined cycle |
| CCS | Carbon capture and sequestration |
| COP | Conference of Parties |
| DAC | Direct air capture |
| DC | Direct current |
| DER | Distributed energy resources |
| DLR | Dynamic line rating |
| EBR | Experimental breeder reactor |
| ERCOT | Electric Reliability Council of Texas |
| EU | European Union |
| EV | Electric vehicle |
| eV | Electron volt |
| FERC | Federal Energy Regulatory Commission |
| $GtCO_2$ | Billion tons of carbon dioxide |
| GW | Gigawatt |
| Hz | Hertz |
| IAEA | International Atomic Energy Agency |
| ICRP | International Commission on Radiation Protection |
| IPCC | Intergovernmental Panel on Climate Change |
| ISO-NE | New England Interconnect |
| Kwh | Kilowatt hour |

| | |
|---|---|
| LCOE | Levelized cost of electricity |
| LCOE | Levelized cost of energy |
| LNT | Linear no threshold |
| LWR | Light water reactor |
| MeV | Million electron volts |
| MISO | Midcontinent Independent System Operator |
| MOX | Mixed oxide fuel |
| mph | Miles per hour |
| mSv | Millisievert |
| NASA | National Aeronautics and Space Administration |
| NRC | Nuclear Regulatory Commission |
| NYISO | New York Independent System Operator |
| O&M | Operations and maintenance |
| OC | Open cycle |
| OECD | Organization for Economic Cooperation and Development |
| PJM | Pennsylvania-New Jersey-Maryland Interconnect |
| PPA | Power purchase agreement |
| PWR | Pressurized water reactor |
| R | Roentgen |
| REC | Renewable energy credit |
| RGGI | Regional Greenhouse Gas Initiative |
| rpm | Revolutions per minute |
| RTO | Regional transmission authority |
| SMR | Small modular reactor |
| SSP | Southwest Power Pool |
| Sv | Sievert |
| TRISO | Tri-structural-isotropic |
| TSO | Transmission system operator |
| NPT | Treaty on the Nonproliferation of Nuclear Weapons |
| U.S. | United States |
| UK | United Kingdom |
| UNSCEAR | United Nations Scientific Committee on the Effects of Atomic Radiation |
| VPP | Virtual power plant |
| VRE | Variable renewable energy |
| yr | Year |

# Bibliography


Allison, Wade. 2009. *Radiation and Reason, the Impact of Science on a Culture of Fear*. York Publishing Services.

Anguin, Meredith. 2020. *Shorting the Grid, the Hidden Fragility of Our Electric Grid*. Carnot Communications.

Blume, Steven W. 2017. *Electric Power System Basics for the Nonelectrical Professional*. IEEE Press/Wiley.

Bradford, Travis. 2018. *The Energy System, Technology, Economics, Markets, and Policy*. MIT Press.

Bryce, Robert. 2020. *A Question of Power, Electricity and the Wealth of Nations*. Public Affairs.

Calabrese, E. J. 2009. The Road to Linearity: Why Linearity at Low Doses Became the Basis for Carcinogen Risk Assessment. *Archives of Toxicology* 83 (3): 203–225.

Carboncredits.com. 2024. *The-ultimate-guide-to-understanding-carbon-credits*.

Cleary, Kathryne, et al. 2021. *FERC 101 Electricity Regulation and the Federal Energy Regulatory Commission*. Resources for the Future.

Danish Wind Industry Association. 2003. *A Wind Energy Pioneer: Charles F. Brush*.

Denholm, Paul, et al. 2010. *The Role of Energy Storage with Renewable Electricity Generation*. NREL/TP-6A2-4718.

Ekins, Paul. 2024. *Stopping Climate Change, Policies for Real Zero*. Routledge.

Federal Energy Regulatory Commission. 2024. *Energy Primer: A Handbook of Energy Market Basics*.

Gates, Bill. 2022. *How to Avoid a Climate Disaster*. Vintage Books.

Goldstein, Joshua S., and Staffan A. Qvist. 2019. *A Bright Future. How Some Countries Have Solved Climate Change and the Rest Can Follow*. New York: Public Affairs.

Griffin, James M., and Steven L. Piller, eds. 2005. *Electricity Deregulation Choices and Challenges*. University of Chicago Press.

Harrison-Atlas, Dylan, et al. 2022. Dynamic Land Use Implications of Rapidly Expanding and Evolving Wind Power Deployment. *Environmental Research Letters* 17:044064.

Herzog, Howard M. 2018. *Carbon Capture*. MIT Press.

Hewlett, Richard G., and Jack M. Holl. 1989. *Atoms for Peace and War 1953–1962, Eisenhower and the Atomic Energy Commission*. University of California Press.

Hogan, John. 1987. *A Spirit Capable: The Story of Commonwealth Edison*. Chicago Review Press.

Houston Chronical Editorial. 2023. *Take the Money and Plug. Orphan Wells Mess in Texas*, January 12.

Jones, Lawrence E., ed. 2017. *Renewable Energy Integration*. 2nd ed. Academic Press/Elsevier.



© The Editor(s) (if applicable) and The Author(s), under exclusive license to Springer Nature Switzerland AG 2026

E. E. Lewis, *Renewables or Nuclear*,

https://doi.org/10.1007/978-3-032-08074-5

Jones, Robert L., Jr. 1994. Cutting the Bill for Commonwealth Edison's Nuclear Power Plants: Important Gains for Illinois Public Utility Customers. *Loyola Consumer Law Review* 6 (2): 3.

Kahneman, Daniel. 2011. *Thinking, Fast and Slow*. New York: Farrar, Straus and Giroux.

Kirschen, Daniel S., and Goran Strbac. 2019. *Fundamentals of Power System Economics*. Wiley.

Langford, R. Everett. 2004. *Introduction to Weapons of Mass Destruction: Radiological, Chemical, and Biological*. John Wiley & Sons.

Larson, E., et al. 2021. *Net-Zero America: Potential Pathways, Infrastructure, and Impacts*. Princeton UP.

Lewis, E. E. 1977. *Nuclear Power Reactor Safety*. Wiley.

Lewis, E. E. 2008. *Fundamentals of Nuclear Reactor Physics*. Academic Press/Elsevier.

Lewis, E. E. 2014. *How Safe Is Safe Enough? Technological Risks, Real and Perceived*. Carrel Books/Simon & Schuster.

Lomborg, Bjorn. 2023. *Best Things First*. Copenhagen Consensus Center.

Long, J. C. S., et al. 2021. Clean Firm Power is the Key to California's Carbon-Free Energy Future. *Issues in Science and Technology*. National Academies Press.

Matthews, Juan, et al. 2024. *The Road to Net Zero: Renewables and Nuclear Working Together*. Dalton Nuclear Institute, University of Manchester.

National Academies. 1996. *Radiation in Medicine: A Need for Regulatory Reform*.

National Academies. 2015a. *Climate Intervention: Carbon Dioxide Removal and Reliable Sequestration*.

National Academies. 2015b. *Climate Intervention: Reflecting Sunlight to Cool Earth*.

National Academies. 2021. *The Future of Electric Power in the United States*.

National Academies. 2023. *Laying the Foundation for New and Advanced Nuclear Reactors in the United States*.

National Academies. 2024a. *A Research Agenda Toward Atmospheric Methane Removal*.

National Academies. 2024b. *Nuclear Terrorism: Assessment of U.S. Strategies to Prevent, Counter, and Respond to Weapons of Mass Destruction*.

National Academies. 2025. *Electricity System Operability and Reliability Under Increasing Complexity*.

OECD. 2012. *Plant-Level, Grid-Level and Total System Costs Nuclear Energy and Renewables: System Effects in Low-carbon Electricity Systems*. NEA No. 7056.

OECD. 2019. *The Costs of Decarbonization: System Costs with High Shares of Nuclear and Renewables*. NEA No. 7299.

OECD. 2021. *Nuclear Energy in the Circular Carbon Economy: A Report to the G20*. NEA No. 7567.

OECD. 2024a. *System Cost Analysis for Integrated Low-Carbon Electricity Systems*. NEA No.7668.

OECD. 2024b. *Small Modular Reactors Dashboard: Second Edition*. NEA No.7671.

Pindyck, Robert S. 2022. *Climate Future: Averting and Adapting to Climate Change*. Oxford University Press.

Shute, Nevil. 1958. *On the Beach*. New York: New American Library.

Smith, Wake. 2022. *Pandora's Toolbox: The Hopes and Hazards of Climate Intervention*. Cambridge University Press.

The Royal Society. 2023. *Large-Scale Electricity Storage*.

Tucker, Colin. 2019. *How to Drive a Nuclear Reactor*. Springer.

U.S. Energy Information Administration. 2022. *Levelized Costs of New Generation. The Annual Energy Outlook*.

Vieira da Rosa, Aldo. 2013. *Fundamentals of Renewable Energy Processes*. 3rd ed. Academic Press/Elsevier.

Von Meier, Alexandra. 2014. *Electric Power Systems, A Conceptual Introduction*. 2nd ed. IEEE Press/Wiley.

Weart, Spencer R. 2012. *The Rise of Nuclear Fear*. Harvard UP.

Wilcox, J., B. Kolosz, and J. Freeman. 2021. *DCR Primer*.

Willrich, Mason. 2017. *Modernizing America's Electricity Infrastructure*. MIT Press.
World Nuclear Association. 2023. *Economics of Nuclear Power*. London: York House.

# Websites

https://hps.org/hpspublications/historylnt/episodeguide.html
https://learn.pjm.com/three-priorities
http://www.virginiaplaces.org/energy/pumpedstorage.html
https://www.osti.gov/opennet/manhattan-project-history/Events/1942-1944_pu/cp-1_critical.htm
https://www.nrdc.org/stories/hundreds-workers-who-cleaned-countrys-worst-coal-ash-spill-are-now-sick-and-dying
https://earthjustice.org/feature/coal-ash-map-sites-legacy-inactive-regulated
https://world-nuclear.org/information-library/plutonium.aspx

# Index